UNSATURATED SOILS

ADVANCES IN TESTING, MODELLING AND ENGINEERING APPLICATIONS

PROCEEDINGS OF THE SECOND INTERNATIONAL WORKSHOP ON UNSATURATED SOILS, 23–25 JUNE 2004, ANACAPRI, ITALY

Unsaturated Soils

Advances in Testing, Modelling and Engineering Applications

Edited by

C. Mancuso
Università degli Studi di Napoli Federico II, Italy

A. Tarantino
Università degli Studi di Trento, Italy

A.A. BALKEMA PUBLISHERS   LEIDEN / LONDON / NEW YORK / PHILADELPHIA / SINGAPORE

Copyright © 2005 Taylor & Francis Group plc, London, UK

All rights reserved. No part of this publication or the information contained herein may be reproduced, stored in a retrieval system, or transmitted in any form or by any means, electronic, mechanical, by photocopying, recording or otherwise, without written prior permission from the publisher.

Although all care is taken to ensure the integrity and quality of this publication and the information herein, no responsibility is assumed by the publishers nor the author for any damage to property or persons as a result of operation or use of this publication and/or the information contained herein.

Published by: A.A. Balkema Publishers, a member of Taylor & Francis Group plc
  www.balkema.nl and www.tandf.co.uk

ISBN 04 1536 742 5

Printed in Great Britain by Antony Rowe Ltd, Chippenham, Wiltshire

Table of Contents

Preface	VII
Acknowledgements	IX
Sponsors	XI

Shear strength behaviour of a reconstituted clayey silt 1
M. Boso, A. Tarantino & L. Mongiovì

Experimental study on the hydro-mechanical behaviour of a silty clay 15
C. Buenfil, E. Romero, A. Lloret & A. Gens

An experimental study on a partially saturated pyroclastic soil: the Pozzolana
Nera from Roma 29
E. Cattoni, M. Cecconi & V. Pane

On the suction and the time dependent behaviour of reservoir chalks of North Sea oilfields 43
G. Priol, V. De Gennaro, P. Delage & Y.-J. Cui

Experimental study on highly compressible neutralised and non neutralised residues
exposed to drying 55
L.F. de Souza Villar & T.M.P. de Campos

Options for modelling hydraulic hysteresis 71
Y.K. Kazimoglu, J.R. McDougall & I.C. Pyrah

Modelling suction increase effects on the fabric of a structured soil 83
A. Koliji, L. Laloui, O. Cuisinier, & L. Vulliet

A bounding surface plasticity model for unsaturated clay and sand 95
A.R. Russell & N. Khalili

Modelling the *THM* behaviour of unsaturated expansive soils using a double-structure
formulation 107
M. Sánchez, A. Gens & S. Olivella

A thermodynamically based model for unsaturated soils: a new framework for
generalized plasticity 121
R. Tamagnini & M. Pastor

Miscellaneous

Opening lecture/Discussion leaders	137
Author addresses	139
List of participants	141

Author index 143

Preface

This volume brings together the contributions to the *Second International Workshop on Unsaturated Soils: Advances in testing, modelling and engineering applications* (Anacapri, Italy – June 22–24, 2004), promoted by the Università di Napoli Federico II (Italy) and the Università di Trento (Italy). Just as the first workshop, held in Trento in 2000, this second one was intended to be a forum for discussing recent advances in unsaturated soil mechanics.

As novices in research very soon learn: sharing ideas, troubles and doubts is essential for moving forward. The spirit of the workshop was to encourage free discussion about unsaturated soil mechanics and to foster the interaction between young and experienced researchers. Ten postgraduate and post-doctoral researchers from different countries were invited to give detailed presentations of their PhD work. The discussion was stimulated by giving enough time for detailed presentations and by inviting experienced researchers in the field of unsaturated soils to lead the discussion sessions. We spent three days in the amazing and relaxed setting of Anacapri, the hilly part of the island of Capri, debating and exchanging ideas, or just talking. We had the chance of getting to know each other and we hope the young researchers will now feel less isolated when struggling with difficulties, experimental errors and numerical results hard to interpret.

We would like to thank the speakers and all the participants for the interest shown in the presentations and the lively discussion. Our gratitude also goes to the discussion leaders, who guided the debate following the presentations, raised searching questions and were generous in their suggestions. We wish to thank Prof. Eduardo Alonso, for giving the opening lecture on *Advances in coupled modelling of embankments and earthdams* and for his inputs to the discussion. Thanks also go to TC6 of the ISSMGE and the Italian Geotechnical Association (AGI) for supporting the workshop.

We hope that the papers published in the proceedings, revised after the comments from the audience and review by the discussion leaders, will interest other researchers in unsaturated soils and will stimulate new ideas for future studies.

C. Mancuso
A. Tarantino

Acknowledgements

We wish to thank Dr. Roberto Vassallo and Ms. Maria Claudia Zingariello for their precious help in the organisation of the workshop.

Sponsors

The editors wish to thank the sponsors
for their financial support to the proceedings of the workshop

MEGARIS s.a.s.

Electronics and electromechanics firm specialised in custom applications
for soil testing, experimental aerodynamics, and biomedical research

Via P. Amato, 39 - 81100 Caserta, Italy
Tel./Fax: 0823/302090
Website: http://www.megaris.it
E-mail: megaris@megaris.it

LAND SERVICE S.c.r.l.

Firm specialised in in-situ geotechnical, geoenvironmental, and
hydrogeological investigations and monitoring.

Via Vittorio Veneto, 26 - 39100 Bolzano, Italy
Tel.: 0471/285434; Fax: 0471/285435
Website: http://www.landservice.it
E-mail: info@landservice.it

Shear strength behaviour of a reconstituted clayey silt

M. Boso, A. Tarantino & L. Mongiovì
Dipartimento di Ingegneria Meccanica e Strutturale, Università degli Studi di Trento, Italy

ABSTRACT: The paper presents results from shearbox tests on a reconstituted unsaturated clayey silt. Tests were carried out using a shearbox with the facility to monitor suctions using high-suction Trento tensiometers. Samples were initially normally consolidated at vertical stress of 100 or 300 kPa and were subsequently air dried. These two consolidation pressures produced unsaturated samples having different void ratio and this made it possible to investigate the influence of void ratio on water retention characteristics and shear strength behaviour. Tests were interpreted in the light of microscopic and macroscopic models recently presented in the literature to account for hydro-mechanical coupling in unsaturated soils.

1 INTRODUCTION

Unsaturated shear strength is of relevance in many geotechnical applications involving either compacted or natural soils (road and railway embankments, flood defences, earth dams, retaining structures, landslides). The unsaturated shear strength can be investigated using the shearbox apparatus. This type of test has both advantages and limitations. The distribution of stresses along the plane of shear is non-uniform, the stress pattern is complex and the directions of the planes of principal stress rotate as the test proceeds. However, test duration is relatively short in comparison with triaxial tests that can take several weeks (if not months) to complete when testing unsaturated soils. In addition, data from triaxial tests are difficult to interpret when a shear band forms within the specimen and the direct shear test permits a better understanding of the post-peak behaviour when compared to triaxial test.

So far, direct shear tests on unsaturated soils have been performed using the axis-translation technique (Fredlund and Morgestern 1978; Escario and Saéz 1986; Gan et al. 1988; Vanapalli et al. 1996, Vaunat et al. 2002). This technique consists in increasing the ambient air pressure to values greater than atmospheric, so as to move the pore water pressures into the positive range. The axis-translation technique is an indirect method of suction measurement and control and its validity has not been fully assessed. As a result, laboratory conditions may not be representative of field conditions, where air pressure is atmospheric and pore water pressure is negative. When using this technique, problems also arise from evaporation of soil water into the air pressure line (Romero 2001) and air diffusion through the high air-entry filter.

Perhaps the main limitation of the axis translation technique is the difficulty of investigating high degrees of saturation. Here the air phase is discontinuous and data are difficult to interpret. As a consequence, the axis-translation technique does not appear suitable for investigating shear strength along the transition from unsaturated to quasi-saturated states.

Wetting paths leading to quasi-saturated conditions are very important in engineering practice, as they are often associated with rainfall triggered landslides. An interesting alternative to axis-translation is the execution of tests under atmospheric conditions with direct measurement of negative pore water pressure. This type of test has made been possible by the use of high-range tensiometers, first developed by Ridley and Burland (1993; 1995).

A shearbox with the facility to monitor suctions was first developed at the University of Trento (Caruso & Tarantino 2004). It incorporated Trento high-range tensiometers (Tarantino and

Table 1. Basic properties of the BCN silt.

Clay (%)	Silt (%)	Sand (%)	G_s	w_h^* (%)	w_L (%)	w_P (%)	$I_P = w_L - w_P$
20	43	37	2.66	1.8	31.8	16.0	15.8

* Hygroscopic humidity (mass basis) at a relative humidity of 50%.

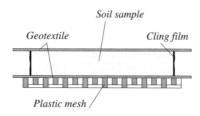

Figure 1. Desaturation by air drying.

Mongiovì 2002; 2003) and was equipped with an anti-evaporation system to carry out tests at constant water content.

Using this shearbox, the shear strength of clayey silt was investigated. Samples were reconstituted from slurry and then air dried. The lack of data on reconstituted unsaturated soils was the main motivation for testing samples in a reconstituted state. Two series of samples were prepared by applying consolidation vertical pressure of 100 and 300 kPa. This produced unsaturated specimens with different void ratio and allowed the investigation of the effect of void ratio and degree of saturation on shear strength. Tests were interpreted in the light of microscopic and macroscopic models recently presented in the literature to account for hydro-mechanical coupling in unsaturated soils.

2 MATERIAL AND SPECIMEN PREPARATION

The material was withdrawn from the Campus Nord of the Universitat Politècnica de Catalunya of Barcelona. Physical properties of the clayey silt are shown in Table 1. The clay fraction is constituted predominantly by illite (Barrera 2002).

To fabricate samples, slurry was prepared at water content two times the liquid limit and then consolidated under one-dimensional condition in a 105 mm diameter consolidometer. Two series of normally consolidated samples were prepared by consolidating the slurry to vertical pressures of 100 and 300 kPa respectively.

After consolidation, samples were air-dried to target water contents estimated by weighing the sample. Evaporation was slowed down to avoid crack formation and to obtain a water content distribution as uniform as possible. To this end a synthetic textile was placed on top and bottom base and a clingfilm was placed on the lateral surface of the sample (Figure 1). When the target water content was reached, samples were sealed into two plastic bags and stored in a high-humidity chamber for moisture equalization for at least one week. After equalization, specimens for shear tests were cut using a square sampler 60 mm side and then trimmed to 10 mm height. Specimens for suction measurement in the airtight suction measurements box were cut using a ring sampler 60 mm diameter and then trimmed to 20 mm height.

3 EXPERIMENTAL PROCEDURES

3.1 Suction measurements on unloaded specimens

Specimens were cut from air-dried samples and placed into an airtight cell. The cell consists of two plates clamped over a ring containing the soil specimen (Figure 2a). Two high-suction tensiometers

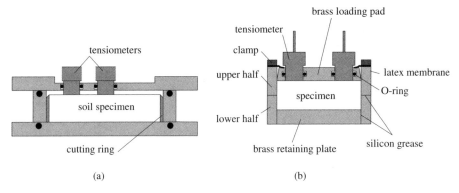

Figure 2. (a) Air-tight suction measurement box. (b) Suction-monitored shearbox.

(Tarantino and Mongiovì 2002) were installed in the upper plate of the cell ensuring air-tightness by means of O-rings.

Measurement lasted sufficient time to allow suction equalization. The volume of the specimen was calculated from the inner diameter and height of the sampler. As a consequence degree of saturation and void ratio could be back calculated in addition to water content. Using this procedure one specimen was required to determine a single water retention datum in terms of suction, water content, and degree of saturation. Water retention curves were therefore determined by testing several samples air-dried to different water contents.

For purpose of comparison, the water retention curve was also determined using a single specimen air-dried in steps and measuring suction at the end of each drying step. To prevent evaporation as much as possible, the specimen was wrapped with several layers of clingfilm. As the specimen could freely shrink, it was not possible to accurately measure specimen dimensions and, hence, its degree of saturation.

3.2 Constant water content shear tests

Constant water content shear tests were performed using a modified shearbox (Figure 2b) which is described by Caruso and Tarantino (2004). After trimming, the specimen was extruded from the sampler and inserted into the shearbox. The loading pad was promptly placed over the specimen to avoid soil-water evaporation. A small vertical pressure (14 kPa) was applied to ensure contact between the loading pad and the specimen and the latex membrane was clamped over the upper half of the shearbox. The tensiometers were inserted into the loading pad and blocked with small caps tightened with three bolts to the loading pad. While tightening the tensiometer caps, the pressure applied to insert the tensiometer was never released to avoid detachment of the tensiometer due to the elastic rebound of the O-ring. The tensiometers were left to equalize for at least one night. The vertical load was increased in steps to reach the final vertical stresses (100, 300, or 500 kPa). The following stress steps were adopted: 50, 100, 200, 300, 400, and 500 kPa. After each pressure increment, suction recorded by the tensiometers and vertical displacement were left to equalize.

After completing the compression stage, specimens were sheared at horizontal displacement rate of 7 mm/day. This rate is seven times greater than the rate that would have been adopted for saturated specimens and was then considered adequate.

4 EXPERIMENTAL PROBLEMS

4.1 The tensiometer paste

A soil paste prepared using the clayey silt fraction of the soil tested ($d < 0.075$ mm) was used to make contact between the tensiometer and the specimen. The water content of the paste may strongly affects the equalization time of tensiometer and also the nature of contact of the tensiometer with the specimen if it is not properly chosen. This is illustrated by the following examples.

Figure 3 shows the compression stage of a shear test where the specimen was loaded in steps to the final vertical stress of 300 kPa. The paste was tentatively prepared at relatively high water content (close to the liquid limit). It was supposed that a wetter paste could better fill any irregularities in the surface of the specimen. At the same time, it was assumed that the paste could ensure contact even if the tensiometer moved backward during installation because of the elastic rebound of the O-ring.

Matric suction was initially let to equalize under a vertical stress of 14 kPa. It may be noticed that the difference in the suction recorded by the two tensiometers is very large. This difference is unusual when using high-suction tensiometers. Moreover suction recorded by the tensiometers is affected by very large fluctuations (150 kPa), again unusual in tensiometer measurements.

As shown in the figure enlargement, these fluctuations were periodical and it was checked that they were in phase with the temperature fluctuations because of the ON/OFF operation of the air-conditioning system ($T = 20 \pm 0.5°C$). As vertical load increased the difference in suction recorded by the two tensiometers reduced and fluctuations dampened.

This tensiometer response can be explained as follows. After installation, the tensiometer paste experienced significant shrinkage as its initial water content was very high. Since the tensiometer was locked in place by the O-ring in the loading pad, the paste partially detached from the specimen surface and cavities formed at the interface between the paste and the specimen (Figure 4a). The partial pressure of water vapour within these cavities was strongly affected by temperature fluctuations even though room air temperature remained within the range $20 \pm 0, 5°C$. Water vapour cyclically condensed and evaporated from the wall of the cavities and accordingly suction cyclically decreased and increased.

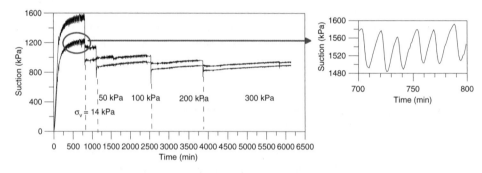

Figure 3. Bad contact of tensiometers (compression stage).

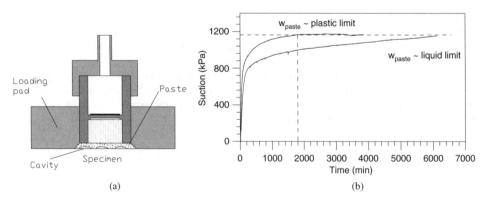

Figure 4. (a) Contact between the tensiometer and the specimen. (b) Suction measurement performed using tensiometer pastes with different water content.

As the vertical load increased, these cavities progressively collapsed and the contact between the paste and the tensiometer improved. Accordingly, tensiometer fluctuations almost vanished and the difference in the tensiometer readings reduced to 50 kPa under the final vertical stress of 300 kPa. This example suggests that tensiometer fluctuations can then be taken as an indicator of bad contact, especially when a single tensiometer is used and it is not possible to check malfunctioning by comparing to independent readings.

Furthermore the water content of the paste may significantly affect the equalization time of suction measurement if not properly chosen. Figure 4b shows two tests performed on specimens with similar suctions and different water content of the paste. The equalization time for the paste having lower water content (approximately the plastic limit) is about of 1800 minutes. The suction measurement performed using the paste with higher water content (approximately the liquid limit) does not seem to reach equalization even after 6000 minutes.

This indicates that the water content of the paste must be kept as low as possible. However contact may not establish if the paste water content is excessively low. Optimal water content must therefore be chosen by trial and error.

5 TEST RESULTS

5.1 *Water retention curves*

Figure 5a shows the retention curve of specimens initially normally consolidated at 100 kPa vertical stress and then air-dried. The suction measurements carried out using the air tight suction measurement box are compared with those performed on a single specimen using the cling film to isolate the sample. At given water content, suction measured on the specimen 'isolated' using the cling film is significantly higher than that measured using the airtight box. This is because water evaporated from the paste that connects the tensiometers with the specimen produced a higher suction than that corresponding to the average water content of the specimen. It is worth noticing that evaporation occurred even if the paste was covered with care (a small hole was made in the cling film and then stretched to allow the insertion of the tensiometer).

Figure 5a emphasises that suction measurement requires (1) isolation of the air surrounding the specimen and (2) equilibrium of the soil water with the surrounding water vapour. If these conditions are not achieved, suction can be significantly overestimated. This error typically affects, for example, suction measurement carried out using the pressure plate, as the specimen is enclosed in a large chamber open to the atmosphere through the air pressure regulator.

Figure 5b shows the retention curves of specimens initially normally consolidated at the vertical stresses of 100 and 300 kPa and then air-dried. Suction measurements were carried using the airtight

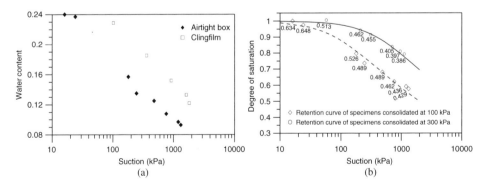

Figure 5. (a) Suction measurements using the airtight suction measurement box and the cling film to isolate the specimen. (b) Water retention curves of specimens initially normally consolidated at vertical stresses of 100 and 300 kPa and then air dried. The void ratio is indicated near the symbol.

cell and the water retention curves are therefore relative specimens under zero total stress. Data were interpolated using Van Genuchten's equation (1980):

$$S_r = \left[\frac{1}{1+(\alpha s)^n}\right]^{1-\frac{1}{n}} \quad (1)$$

where S_r is the degree of saturation, s the suction, α and n model parameters.

The two retention curves differ significantly and this difference is associated with the different void ratio of the two series of specimens (void ratios are indicated in the figure near the symbols). Specimens initially consolidated at 300 kPa vertical stress have lower void ratios and this shifts the air-entry suction to higher values. As a result, the degrees of saturation are higher than those of the specimens consolidated at 100 kPa vertical stress.

The dependency of the water retention curve on void ratio has been previously observed (Romero and Vaunat 2000; Karube and Kawai 2001). However, the most remarkable aspect of the data shown in Figure 5b is that relatively small variations in void ratio significantly modified the position of the water retention curve. In other words, the water retention curve of the soil tested appears to be very sensitive to changes in void ratio.

5.2 Shear tests

Shear tests at constant water content were performed at vertical stresses of 100, 300, and 500 kPa on specimens initially consolidated at 100 or 300 kPa vertical stress.

Figure 6 shows two shear tests performed on two specimens having different water contents but sheared at the same vertical stress (300 kPa). Both specimens were initially consolidated at 300 kPa vertical stress and air dried. In the figure, the average degree of saturation, the shear strength, the vertical displacement and suction are plotted versus the horizontal displacements. The specimen with the higher water content (Figure 6a) reached saturation during compression and remained saturated during the subsequent shearing. For this specimen, suction initially decreased (pore-water pressure increased) and then levelled off at zero suction. As the specimen is surrounded by air, pore-water pressure could not attain positive values. To maintain zero porewater pressure, a small amount of water had to extrude through the gap between the loading pad and the upper half. This extrusion was accompanied by little loss of soil and this explains the non-zero vertical displacements. The degree of saturation was calculated assuming that the mass of the soil solids remained constant during the test. However this assumption is no longer valid when extrusion occurs and this explains why the figure shows a calculated degree of saturation greater than one. This test at zero suction and unit degree of saturation shows that it was possible to investigate the transition from unsaturated to saturated states.

The specimen with the lower water content exhibits a higher suction at the end of the compression stage (Figure 6b). The shear strength shows a peak which is consistent with the dilatant behaviour of the specimen. This 'overconsolidated' type of behaviour is likely associated with the suction-induced hardening occurring during the air-drying process (suction reached 1200 kPa at the end of the air-drying stage).

During shearing the suction decreased reaching 500 kPa and then levelled off at about 4 mm horizontal displacement. In spite of the dilatant behaviour, suction never increased, as would intuitively be expected considering that water content was constant and hence degree of saturation decreased.

This behaviour was common to all tests where dilatant behaviour was observed and is illustrated in Figure 7a where degree of saturation-suction paths are reported for the compression and shearing stage of specimens initially consolidated at 300 kPa vertical stress and sheared at 300 kPa vertical stress. At shearing, the specimen first exhibit contractile behaviour (the degree of saturation increases) and then dilates (the degree of saturation decreases). However, suction always decreases even when the specimen starts dilating. An interpretation of this apparently surprising behaviour will be given later in this paper by invoking the void ratio dependency of the main drying curve.

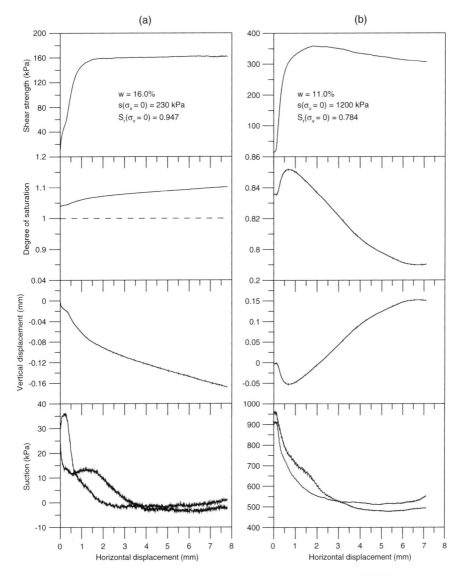

Figure 6. Shear tests on specimens initially consolidated at the vertical stress of 300 kPa and air-dried at different water contents. The vertical stress during shearing is 300 kPa.

Figure 7b shows the degree of saturation-suction paths for specimens initially consolidated at 100 kPa vertical stress and sheared at 500 kPa vertical stress. At low suctions, the soil exhibits compressive behaviour at shearing and degree of saturation constantly increases. However, it can be noticed a sharp increase in the slope of the suction paths after the compression stage. An interpretation of this behaviour will also be attempted later in the paper.

5.3 Shear strength envelopes

The envelopes of the shear strength versus suction for specimens initially consolidated at the vertical stress of 100 kPa and sheared at 100, 300, and 500 kPa vertical stress are shown in Figure 8a. The saturated envelopes (i.e. the shear strength for the case where the soil remained saturated under any suction) are also reported in the figure. These were obtained from the 'saturated' parameters

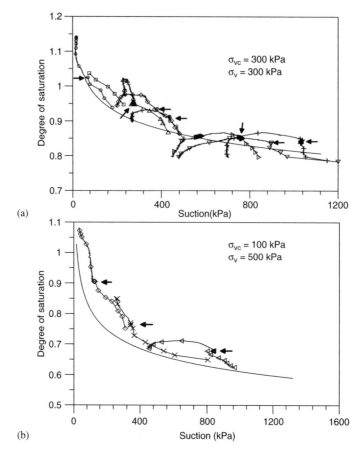

Figure 7. Suction paths during the compression and shearing stage. (a) Specimens initially consolidated at the vertical stress of 300 kPa and sheared at 300 kPa vertical stress. (b) Specimens initially consolidated at the vertical stress of 100 kPa and sheared at 500 kPa vertical stress. The arrows indicate the start of the shearing stage.

determined from shearbox tests on normally consolidated samples kept saturated by immersion in free water.

The shear strength of the specimen sheared at 100 kPa vertical stress that reached zero suction during shearing is very close to the shear strength of the specimens tested under saturated conditions. In other words, the shear strength of the air-immersed sample at zero suction equals the strength of the water-immersed sample at zero water pressure. This result is only apparently trivial. There is little experimental evidence showing the continuity of shear strength behaviour from unsaturated to saturated states.

Inspection of Figure 8a also reveals that the unsaturated envelope detaches from the saturated envelope at suctions that increase as the applied vertical stress increases (as indicated by the arrows in the figure). This can be explained by the coupling between hydraulic and mechanical behaviour. As the vertical stress increases, the void ratio decreases and, hence, the air-entry (or air-occlusion) suction increases. The degree of saturation remain close to unity over a wide range of suction and, accordingly, the shear strength remains close to the saturated value over a wider range of suction.

This result emphasizes that suction and net stress alone are not sufficient to model shear strength and a term related to the degree of saturation should be included in any shear strength criterion to account for hydro-mechanical coupling.

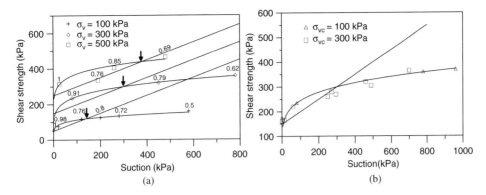

Figure 8. (a) Shear strength versus suction envelopes of specimens initially consolidated at the vertical stress of 100 kPa and sheared at 100, 300, and 500 kPa vertical stress. The degree of saturation is indicated near the symbols. (b) Shear strength versus suction envelopes of specimens sheared at 300 kPa vertical stress.

The most striking aspect shown in Figure is the shear strength of the unsaturated soil at very low suction (high degrees of saturation), which appears to be greater than that of the saturated soil at the same suction. This concerns specimens having a high degree of saturation, in the range 0.85–1.0 where the air-phase is expected to be discontinuous within the pore space. There are two exceptions, the specimen sheared at 500 kPa vertical stress having an average degree of saturation of 0.76 and the specimen sheared at 100 vertical stress having an average degree of saturation of 0.76. These average degrees of saturation are not consistent with that of the other specimens and are probably affected by an error in the measurement of the vertical displacement.

Unsaturated shear strength greater than that of saturated soil at the same suction was also observed in specimens initially consolidated at 300 kPa. As shown in Figure 8b, the unsaturated envelope is positioned above the saturated envelope in the range of low suctions (high degrees of saturation) and crosses the saturated envelope at a suction that is expected to mark the transition from discontinuous to continuous air phase.

6 DATA INTERPRETATION

6.1 *Shear strength of an ideal soil*

Experimental data have shown that shear strength of quasi-saturated soils (discontinuous airphase) is greater than that of the saturated soil at the same suction. This result was unexpected and an attempt was made to provide a theoretical justification of this behaviour.

Let us first consider two spherical rigid particles. If the space between the particles is completely filled with water subject to a suction s (Figure 9a), the intergranular stress σ_i^{sat} equals the suction s.

If a meniscus forms at the interparticle contact (Figure 9b), suction is given by:

$$s = t\left(\frac{1}{c} - \frac{1}{b}\right) \qquad (2)$$

where r is the particle radius, t the surface tension, b the radius of the neck of fluid connecting the two spheres, c the radius of the meridian curve. The intergranular stress is given by:

$$\sigma_i^{meniscus} = \frac{N}{A} = \frac{\pi b^2 t\left(\frac{1}{c} - \frac{1}{b}\right) + 2\pi b t}{\pi r^2} \qquad (3)$$

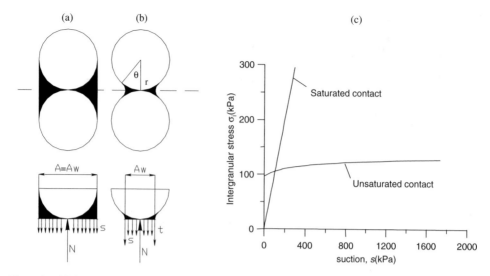

Figure 9. (a) Saturated conditions and (b) Interaction between spherical particles in unsaturated conditions. (c) Intergranular stress versus suction for saturated and unsaturated contact.

where N is the normal force at the interparticle contact and A the total area. Equation (3) reduces to (Fisher 1926):

$$\sigma_i^{meniscus} = \frac{N}{A} = \frac{\frac{2\pi r t}{1 + \tan(\theta/2)}}{\pi r^2} \quad (4)$$

where θ is the angle defining the position of the meniscus junction.

As an example, let us consider the case of $r = 10^{-6}$ m and $\theta = 53°$. Under these conditions, suction is zero ($b = c$) but the intergranular stress σ_i^{unsat} is positive. In other words, the intergranular stress of the unsaturated contact σ_i^{unsat} is greater than that of the saturated contact ($\sigma_i^{sat} \equiv s = 0$).

The intergranular stress for the case of the unsaturated contact is plotted versus suction in Figure 10c, where it is compared with the intergranular stress of the saturated contact. The two curves cross each other and it is interesting to notice that is equivalent to Figure 8 from a qualitative point of view.

Consider now an ideal soil consisting of rigid spheres of equal diameter. If the packing is open, each sphere touches 6 other spheres and the points of contact are the centres of the 6 faces of a cube. Let us assume that the air-water interface is continuous at the boundary of the packing and, hence, the air outside the packing cannot enter the pore space. If suction increases, cavitation will occur in some internal pores, and menisci will form at some interparticle contact. This mechanism of desaturation may be realistic for a quasi-saturated soil where air phase is discontinuous.

Figure 10 shows a scheme of a quasi-saturated packing. The average intergranular stress is given by:

$$\sigma_i^{unsat} = \sigma_i^{bulk} \frac{A_{wb}}{A_{tot}} + \sigma_i^{meniscus} \left(1 - \frac{A_{wb}}{A_{tot}}\right) \quad (5)$$

where σ_i^{bulk} bulk is the intergranular stress in the bulk water region ($\sigma_i^{bulk} = s$), $\sigma_i^{meniscus}$ is intergranular stress in the meniscus water region (given by equation (4)), A_{tot} the total cross area, A_{wb} area occupied by the bulk water.

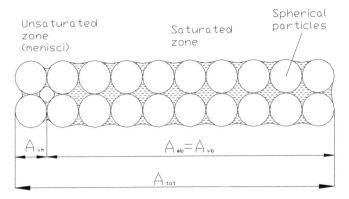

Figure 10. Idealised quasi-saturated soil.

Let us assume that:

$$\frac{A_{wb}}{A_{tot}} = \frac{V_{wb}}{V_v} \qquad (6)$$

where V_v is the volume of voids and V_{wb} the volume of voids occupied by the bulk water. Let us indicate with S_{rm} the degree of saturation of the region occupied by the menisci alone:

$$S_{rm} = \frac{V_{wm}}{V_{vm}} \qquad (7)$$

where V_{wm} is the volume of water of the menisci and V_{vm} is the volume of the pores of the meniscus region. The value of S_{rm} is provided by Fisher (1926) as a function of suction.

It can be shown that

$$\frac{V_{wb}}{V_v} = \frac{S_r - S_{rm}}{1 - S_{rm}} \qquad (8)$$

where S_r is the overall degree of saturation. By combining equations (5), (6), and (8) we obtain:

$$\sigma_i = s \cdot \frac{S_r - S_{rm}}{1 - S_{rm}} + \sigma_i^{meniscus}\left(1 - \frac{S_r - S_{rm}}{1 - S_{rm}}\right) \qquad (9)$$

Consider the water retention curves shown in Figure 11a. The curves are plotted in the range 0.85–1.0 of the degree of saturation which may correspond to the range of discontinuous air phase. If the relationship $S_r = S_r(s)$ is introduced in equation (9), it is possible to determine the intergranular stress σ_i in the quasi-saturated soil and compare it with the intergranular stress in the saturated packing at the same suction. The result of this analysis is shown in Figure 11b. The water retention curve with the lower air-entry suction gives the maximum difference between the intergranular stress of the quasi-saturated packing (σ_i) and the of the saturated packing (s). The difference σ_i-s tends to vanish as the air-entry suction increases.

In real soils, it may be therefore possible that particular water retention curves gives, in the range of high degrees of saturation, an intergranular stress greater than that of the saturated soil at the same suction.

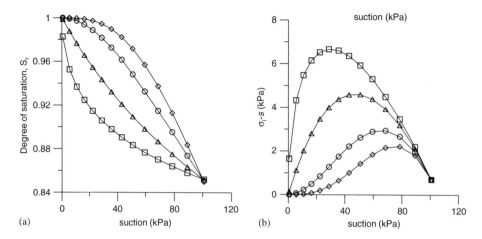

Figure 11. Effect of water retention curve on intergranular stress in quasi-saturated and saturated packing. (a) Water retention curves. (b) Difference of intergranular stresses between quasi-saturated (σ_i) and saturated (s) packing.

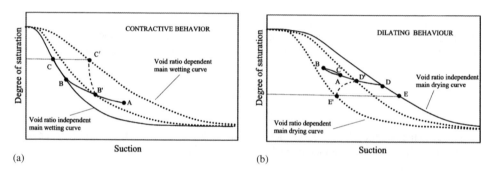

Figure 12. Conceptual model for degree of saturation versus suction paths at shearing. (a) Contractile behaviour. (b) Dilatant behaviour.

6.2 *Degree of saturation-suction paths at shearing in deformable soils*

Romero and Vaunat (2000) assumed the existence of a main wetting curve and a main drying curve at constant void ratio that delimits a domain of attainable states. They showed that a variation of void ratio imply variations of the air-entry (or air-occlusion) value and the consequent movement of this domain.

Moving from this concept, it is possible to explain, from a qualitative standpoint, the degree of saturation-suction paths observed at shearing. After consolidation, specimens were air-dried starting from saturated condition. Specimens then moved along main drying curves. Then specimens were compressed in steps at constant water content. During compression the void ratio decreased and consequently the degree of saturation increased. As a consequence, the specimen state left the main drying curve to follow a scanning curve. At the end of the compression stage the soil state was in A inside the hysterisis domain (Figure 12).

Let us focus on the contractile behaviour and, for sake of simplicity, let us first consider the case of a void ratio-independent water retention curve. If volume decreases, the degree of saturation increases and the soil follows a scanning curve, possibly reaching the main wetting curve (path A, B, C in Figure 12a).

Consider the case of a deformable soil, in the sense that the main wetting curve is void ratio-dependent. Initially soil state moves along a scanning curve. As volume start decreasing, the main wetting curve moves rightwards until it meets the soil state in point B′. Then, a further volume decrease causes a displacement of the main wetting curve that drags the soil state upwards (point C′). In fact, the main wetting curve represents the lower bound of the domain of attainable states. In this case, the degree of saturation-suction path presents a sudden change in slope in point B′. This would explain the sudden change in slope of the suction-degree of saturation path observed in Figure 7.

Let us consider the case of a contractile-dilatant behaviour and that of a void ratio-independent main drying curve. When the specimen is compressed, void ratio decreases and the degree of saturation increases (point B). When dilation commences, the soil moves backward following the same scanning curve until it reaches the main drying curve (point D). If the volume continues increasing the soil state would follow the main drying curve until the point E. As a result, for the case of a void ratio-independent main drying curve, suction would always increase during dilation.

Now let us consider a deformable soil where the main drying curve is void ratio-dependent (Figure 12b). Initially, the soil follows the scanning curve to point B. When dilation commences, the path is reversed to point C≡A. At the same time, the volume increase causes the leftwards movement of the main drying curve until it meets the soil state in D′. As volume continues increasing, the main drying curve drags the soil to point E′. As a result, the soil experiences a decrease in suction even though volume increases.

7 CONCLUSIONS

The experimental investigation of the shear strength of a reconstituted unsaturated clayey silt has shown that:

(1) the shear strength of the air-immersed samples at zero suction is equal to the shear strength of the water-immersed samples at zero pore water pressure;
(2) at high degrees of saturation, the shear strength of the quasi saturated soil may be greater than that of the saturated soil at the same suction;
(3) the unsaturated envelope detaches from the saturated envelope at suction that are greater as void ratio decreases. This is due to the increase in the air-entry suction and, more in general, to the dependency of the water retention curve on void ratio. This result shows that any shear strength criterion for deformable soils cannot be represented in terms of net stress and suction alone but should include a term related to the degree of saturation.
(4) at shearing suction always decreased even when the soil exhibited dilatant behaviour and, hence, the degree of saturation increased.

An attempt has been made to justify some of the observed behaviours from a conceptual standpoint. The increase in shear strength of the quasi-saturated soils with respect to that of the saturated soil at the same suction may be explained by the appearance of menisci in the pores where cavitation occurs. The additional shear strength would be therefore due to the surface tension transmitted by the air-water interface.

The decrease in suction observed at shearing in concomitance of dilatant behaviour would be explained by the dependency of the water retention curve on void ratio. As the soil dilates and, hence, its void ratio increases, the main drying curve would move backward so as to drag the soil state to lower suctions.

ACKNOWLEDGEMENTS

The authors wish to thank Marco Bragagna, Senior Technician of the Geotechnical Laboratory for the invaluable help and Sara Tombolato for the helpful discussions.

REFERENCES

Barrera, M. (2002). Estudio experimental del comportamiento hidromecánico de suelos colapsables. PhD Thesis, Universitat Politécnica de Catalunya, Barcelona, Spain.

Caruso, A. & Tarantino, A. 2004. A shearbox for testing unsaturated soils from medium to high degrees of saturation. *Géotechnique*, **54**, No. 4, 281–284.

Escario, V. & Saez, J. 1986. The shear strength of partly saturated soils. *Géotechnique*, **36**, No. 3, 453–456.

Fisher, R.A. 1926. On the capillary forces in an ideal soil; correction of formulae given by W.B.Haines – *Jour. Agr. Sci.*, 16, 492–505.

Fredlund, D. G. & Morgestern, N. R. 1978. Shear strength of unsaturated soils. *Canadian Geotechnical Journal*, **15**, 313–321.

Gan, J.K.M., Fredlund, D.G. & Rahardjo H. 1988. Determination of shear strength parameters of an unsaturated soil using the direct shear test. *Canadian Geotechnical Journal*, **25**, 500–510.

Karube, D. & Kawai, K. 2001. The role of pore water in the mechanical behaviour of unsaturated soils. *Geotechnical and Geological Engineering*, **19**, 211–241.

Ridley, A.M. & Burland, J.B. 1993. A new instrument for the measurement of soil moisture suction. *Géotechnique*, **43**, No. 2, 321–324.

Ridley, A.M. & Burland, J.B. 1995. Mesasurement of suction in materials which swell. *Applied Mechanics Reviews*, **48**, No. 10, 727–732.

Romero, E., Controlled-suction techniques 2001. *4° Simpósio Brasileiro de Solos Nâo Saturados Ñ SAT'2001*, W.Y.Y. Gehling & F. Schnaid (eds), PortoAlegre, Brasil, pp. 535–542.

Romero, E. & Vaunat, J. 2000. Retention curves in deformable clays. In *Experimental Evidence and Theoretical Approaches in Unsaturated Soils*, A. Tarantino & C. Mancuso (eds), pp. 91–106, Rotterdam, A.A. Balkema.

Tarantino, A. & L. Mongiovì, 2000. A study of the efficiency of semipermeable membranes in controlling soil matrix suction using the osmotic technique. *Proceedings of the Asian Conference on Unsaturated Soils*, 18–19 May 2000, Singapore: 303–308. Rotterdam: A.A. Balkema.

Tarantino, A. & Mongiovì, L. 2002. Design and construction of a tensiometer for direct measurement of matric suction. In *Proceedings 3rd International Conference on Unsaturated Soils*, J.F.T. Jucá, T.M.P. de Campos & F.A.M. Marinho (eds.), Recife **1**, 319–324.

Tarantino, A. 2004. Panel Report: Direct measurement of soil water tension. Proc. *3rd Int. Conf. on Unsaturated Soils*, Recife, Brasil, **3**: 1005–1017.

Tarantino, A. & Mongiovì, L. 2003. Calibration of tensiometer for direct measurement of matric suction. *Géotechnique*, **53**, No. 1, 137–141.

van Genuchten, M.Th. 1980. A closed-form equation for predicting the hydraulic conductivity of unsaturated soils. *Soil Sci. Soc. Am. J.* 44: 892–898.

Vanapalli, S.K., Fredlund, D.G., Pufahl, D.E. & Clifton, A.W. 1996. Model for the prediction of shear strength with respect of soil suction. *Canadian Geotechnical Journal*, **33**, 379–392.

Vaunat, J., Romero, E., Marchi, C. & Jommi, C. 2002. Modeling the shear strength of unsaturated soils. *Proceedings 3rd International Conference on Unsaturated Soils*, Recife, Brasil, **1**, pp. 245–251.

Experimental study on the hydro-mechanical behaviour of a silty clay

C. Buenfil, E. Romero, A. Lloret & A. Gens
Departament d'Enginyeria del Terreny, Cartogràfica i Geofísica, Universitat Politècnica de Catalunya, Barcelona, Spain

ABSTRACT: This paper reports an experimental programme aimed at studying the coupled hydro-mechanical behaviour of an unsaturated low-density silty clay. A poorly statically compacted condition was selected to induce an appreciable change of the void ratio by compression and to study its consequences on the water retention properties of the soil. As preliminary work, the characterization of the microstructure of two samples compacted at two different dry densities was performed by mercury intrusion porosimetry and ESEM image analysis to understand the structural changes induced by compression paths. The first dry density corresponds to the initial state of the soil (e = 0.82), whereas the second one (e = 0.55) corresponds to the dry density that reaches the sample after the loading path. In addition, the retention curves of the clay compacted at the two aforementioned dry densities were determined.

Different stress paths under isotropic conditions have been followed to cover a wide range of partially saturated behavioral features using a fully instrumented and automated triaxial cell. The triaxial cell, which is described in the paper, uses axis translation and negative water column techniques to apply suction. The cell, which is an improved version of the cell presented by Romero et al. (1997), has been updated using stepper-motors and a control system to carry out complex and continuous stress paths. In addition, an important effort has been devoted to accurately determine water content changes using an instrumented burette with very sensitive differential pressure transducers. The cell uses local instrumentation (axial LVDTs and mobile radial electro-optical systems) to accurately detect non-uniformity radial strain patterns along the sample height and determine global degree of saturation changes.

The paper describes and presents the relevant results of the different stress paths followed. The open structure of the material displayed important water content changes in the low-suction range when the sample was submitted to wetting and loading. Test results are interpreted within the framework of an elastoplastic model (Alonso et al. 1990) and bounding retention curves (Vaunat et al. 2000), which cope with the coupled hydro-mechanical response of unsaturated soils.

1 INTRODUCTION

Few experimental studies have been focused on the hydro-mechanical coupled behaviour of unsaturated soils when general stress paths are applied (Rampino et al. 2000, Romero 1999, Romero & Vaunat 2000, Barrera 2002). Traditionally, the abundant studies concerning the behaviour during compression under isotropic conditions have been mainly focused on mechanical aspects, such as compressibility variation and yield properties at different suction levels (Alonso et al. 1990, Sivakumar 1993, Rampino et al. 1999, 2000, Chen et al. 1999). Well-posed experimental techniques, as well as accurate testing cells with local instrumentation, are required to obtain reliable results of this coupled response (Romero et al. 1997, Rampino et al. 1999, Barrera 2002).

Clayey soils when compacted on the dry side display a clear double structure formed by clay aggregations. In these soils, two structural levels can be considered: a microstructure inside the aggregates and a macrostructure constituted by the arrangement of aggregates and interaggregates pores. Water in aggregated structures containing micro and macropores is assumed to be retained by capillary effects (free water) and water adsorption mechanisms (adsorbed water) (Barbour 1998, Romero et al. 1999, Vanapalli et al. 1999). Inside micropores of the aggregates, where adsorbed water is predominant, the water content is unaffected by mechanical effects, and inside macropores, where free water is predominant, the water content is sensitive to mechanical actions (Romero et al. 1999).

Water retention characteristics of compacted soils are largely influenced by soil fabric. When this fabric changes due to stress action, the coupling between the mechanical and hydraulic may be perceptible. The low-suction range of the water retention curve as a function of the gravimetric water content is highly dependent on void ratio. The changes of this volumetric variable induced by mechanical actions affect mainly the water storage capacity of the soil at saturation, the air-entry value on drying and the air-occlusion value on wetting (Romero & Vaunat 2000, Karube & Kawai 2001).

Usually, it is considered that when the overall water content is maintained, an increase in confining stress results in an increase of the degree of saturation and, consequently, in a decrease in soil suction. However, in some cases, suction increases have been observed when the soil porosity is reduced due to stress application. Kawai et al. (2003), using measurements of water and air pressure and Tombolato et al. (2003) using tensiometers observed this phenomenon in tests where volume changes due to stress increase were significant. These results were explained considering the changes in water retention characteristics associated with pore volume variation. Unfortunately, the aforementioned authors did not include information about the characteristics of the soil fabric changes nor of the water retention curves of the soil. In order to reach a correct interpretation of the hydro-mechanical soil behaviour during the compaction process and the subsequent application of general stress paths, it is necessary to relate fabric changes to changes in water retention curves. In 1940, Childs already observed that retention curves can be considered as a complementary part in a mechanical analysis since they provide information on the pore size distribution.

The paper contains some results of a laboratory investigation performed on low-density silty clay with the aim of studying the hydraulic response on isotropic loading at low suctions. The results obtained in controlled-suction compression tests are studied jointly with the fabric characterization defined by mercury intrusion porosimetry (MIP) and microphotographs obtained in an environmental scanning electron microscope (ESEM). In addition, the water retention curves of the soil compacted at two different dry densities have been determined in order to study the effects of the porosity changes in these curves.

The results of this experimental study provide a consistent picture of the coupled hydromechanical response of the soil, in which the loading paths clearly affect the shape of the water retention curves and the consequent hydraulic response of the soil.

2 CHARACTERIZATION OF THE TESTED SOIL

2.1 *Tested material and compaction procedures*

Laboratory tests were performed on low plasticity silty clay from Barcelona. This material has a liquid limit of $w_L = 28\%$, a plastic limit of $w_P = 19\%$, a clay-size fraction $\leq 2\,\mu m$ of 19%, a silty fraction of 47% and unit weight of the solids of $\gamma_s = 26.6\,kN/m^3$. The dominant mineral of the clay fraction is illite (Barrera 2002). The hygroscopic water content of the soil at laboratory conditions (relative humidity 47%) is about 1.7%.

Triaxial samples (38 mm diameter and 76 mm high) at a prescribed water content of 12.0% were prepared at a dry unit weight of $\gamma_d = 14.9\,kN/m^3$ (degree of saturation of $Sr = 40\%$), by one-dimensional static compaction under constant water content of 12% and at a constant piston displacement rate of 0.2 mm/min. Maximum fabrication vertical net stress was 0.27 MPa. Lateral stresses were measured by an active lateral stress system, resulting in a lateral stress coefficient at rest of $K_0 = 0.48$. Suction after compaction, $s = 270\,kPa$, was measured using a high-range tensiometer (Ridley & Burland 1993).

Table 1. After compaction properties of samples used for fabric description.

γ_d kN/m³	e	w (%)	Sr (%)	$\sigma_v - u_a$ (MPa)	$p - u_a$ (MPa)
14.9	0.82	12	40	0.27	0.18
16.9–17.1	0.55–0.57	12	57–59	1.0–1.2	0.67–0.8

In Table 1 the mean net stress, $(p - u_a)$, has been estimated from vertical net stress, $(\sigma_v - u_a)$, using $K_0 = 0.48$.

A relatively low dry unit weight was selected with the aim of inducing an open structure, which was susceptible to undergo important void ratio changes on loading and in turn to induce important changes of the water retention properties of the soil.

2.2 Fabric description for two different soil structures

Two samples with quite different packing were prepared in order to observe the influence of fabric on the soil–water interaction properties and the hydro-mechanical behaviour under isotropic loading. Table 1 shows the properties of both types of samples after the compaction process. The larger void ratio is representative of the as-compacted state of a sample with an open structure and the lower value corresponds to the state reached by this sample after undergoing an important compression by loading.

2.2.1 Mercury intrusion porosimetry studies

Pore size distributions (PSD) of statically compacted clayey silt were obtained by mercury intrusion porosimetry (MIP) in order to examine the fabric and pore morphology of the soil with the two different dry densities indicated in Table 1. MIP tests were performed following ASTM D4404 testing procedure in an 'Autopore IV 9500' porosimeter with a mercury intrusion pressure up to 228 MPa (access radii between 360 μm and 5 nm).

Before mercury intrusion, cubic specimens measuring 10 mm on each side were trimmed and freeze dried to remove the pore water. It was assumed that no shrinkage occurred on drying the samples using the freeze drying process.

Figure 1 shows the measured pore size density functions for two samples compacted at very different values of dry unit weight $\gamma_d = 14.9$ kN/m³ and 17 kN/m³. It can be observed that the pore size distribution is clearly bi-modal, which is characteristic of this type of materials (Delage et al. 1996). The dominant values are 455 nm that would correspond to the pores inside clay aggregates and a larger pore size that depends on the compaction dry density and ranges from 19 μm (for $\gamma_d = 17$ kN/m³) and 60 μm (for $\gamma_d = 14.9$ kN/m³). These larger voids would correspond to the inter-aggregate pores. The boundary between the two pore size families can be seen to be around 5 μm, as pores smaller than this magnitude do not appear to be affected by the magnitude of the compaction load. As Figure 1 clearly shows, compaction affects only the pore structure of the larger inter-aggregate pores.

2.2.2 Scanning electron microscope observations

Figure 2 shows ESEM microphotographs of the samples corresponding to the two packing indicated in Table 1. Samples were prepared following the same compaction procedure described before, but freeze drying process was not necessary in this case. The fabrics observed using the microscope are reasonably consistent with PSDs curves. The photographs, taken with a magnification of 250x for the low-density soil and 200x for the high-density soil, clearly show the different sizes of the inter-aggregate pores of the two samples. In the photograph of the soil with low density (Fig. 2a) it is possible to detect large inter-aggregate pores with dimensions between 20 μm and 100 μm, which are consistent with the large pore mode obtained by MIP test. In the high density packing (Fig. 2b) it can be observed a significant decrease in the size of the inter-aggregate pores that display

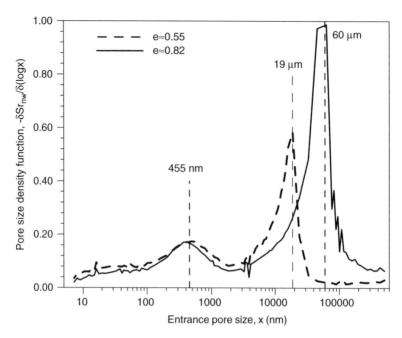

Figure 1. Pore size density function evaluated from MIP results.

Figure 2. Scanning electron micrographs of the compacted silt at (a) lower packing, (b) higher packing.

diameters smaller than 30 μm. The intra-aggregate pores that are visible at the magnification used in the figure present similar sizes for the two packing investigated.

Similar patterns to the ones observed in the PSD and ESEM tests were also obtained by Delage et al. (1996) and Romero et al. (1999) in soils compacted on the dry side of the compaction curve.

2.3 *Water retention curves*

Retention curves of the clayey silt were obtained using a controlled-suction oedometer cell. Samples (50 mm diameter and 10 mm high) were prepared following the same compaction procedure described before in order to obtain the two contrasting packing shown in Table 1.

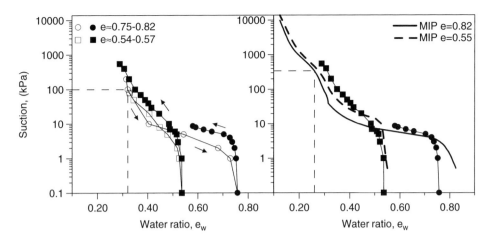

Figure 3. Wetting and drying retention curves on a water ratio basis of the clayey silt at two contrasting void ratios compared to MIP results for different packings at constant porosity.

Axis translation technique with a constant air pressure was used to apply suctions ranging between 550 kPa and a value close to zero. The water pressure was applied by a GDS Instruments Ltd. pressure/volume controller connected to a high air-entry value ceramic (HAEV of 1.5 MPa), which allowed the measurement of water volume changes. Suctions lower than 8 kPa were controlled by placing the pressure controller below the oedometer level (negative water column). The retention curves were obtained at a constant vertical net stress, $(\sigma_v - u_a = 20\,\text{kPa})$, following multi-stage wetting and subsequent drying paths. At this stress level, the loose sample developed some collapse on wetting. Nevertheless, the void ratios during hydration and subsequent drying did not change significantly and the data could be considered as representative of constant volume retention curves.

The relationships between suction and water content in wetting and drying paths for both densities have been plotted in Figure 3, in terms of water ratio e_w (e_w = water volume/solid particles volume). The water ratio $e_w = G_s w$ was considered as the work conjugate volumetric variable associated with suction, in the same way as the volumetric variable void ratio e was associated with the net stress variable (Romero & Vaunat 2000). As observed, the wetting and drying retention curves are sensitive to void ratio value in the small range of suctions. The water content at null suction depends on the void ratio and therefore the curves markedly diverge to reach the water contents corresponding to their saturated conditions. Moreover, the changes in void ratio also affect the air-entry value on drying. The low-density soil has a lower air-entry value (5 kPa) compared to the high-density soil (10 kPa). Void ratio effects are also detected in the crossing of the two wetting branches, as a consequence of their influence on the air-occlusion value (water-entry value) of the curves. Differences in the air-entry value in the two packing can be directly related to the differences in the size of the inter-aggregate pores that have been observed in the MIP results.

MIP data can be used to obtain water retention curves. Romero et al. (1999), presented a methodology to obtain matric suction – saturation relationships taking into account that the mercury intrusion can be considered as an equivalent process to water desaturation by gradually increasing external air pressure (no wetting fluid). Therefore, mercury intrusion with contact angle $\theta_{nw} = 140°$ and surface tension $\sigma_{Hg} = 0.484\,\text{N m}^{-1}$ is comparable with soil dehydration by air injection with contact angle $\theta_{nw} = 180°$ and water surface tension $\sigma_{Hg} = 0.072\,\text{N m}^{-1}$. Water degree of saturation, Sr, corresponding to the 'main drying curve' obtained by air overpressure should be related to the volume the pores not intruded by mercury, $Sr = 1 - Sr_{nw}$. The nonwetting mercury degree of saturation, $Sr_{nw} = n/n_o$, is the introduced porosity n normalised by the total porosity n_o. Finally, in order to take into account the strongly adsorbed water corresponding to the residual water content, w_r, the following expression was proposed: $Sr = (1 - Sr_{nw}) + w_r\,Sr_{nw}/w_{sat}$, where w_{sat} is the gravimetric water content in saturated state.

Figure 3 shows the comparison between the retention curves obtained using the suction controlled oedometer cell and the curves derived from MIP results. The two types of curves display similar results.

From oedometer test results, it can be observed that the water retained at suctions higher than 100 kPa is independent of the total void ratio (or dry density). It seems plausible that the water retained at these high suctions (water content about 11.8%, $e_w = 0.33$) belongs to the intra-aggregates pores where the total porosity plays no relevant role (Romero et al. 1999). From MIP tests, this borderline suction may be situated at 340 kPa, which corresponds to a water content of about 9.6 % ($e_w = 0.26$).

Concerning the shape of the suction curves, following the same nomenclature of the hydro-mechanical model proposed by Vaunat et al. (2000), on first wetting before compaction, the soil follows the 'main wetting curve', which acts as a state boundary curve in the w:s plane. This 'main wetting curve' changes on subsequent compaction at constant water content, due to its dependence on void ratio. The as-compacted state after this compression process is not located on this 'main wetting curve'. In this way, a subsequent wetting phase will follow a 'scanning curve' with a slope steeper than the slope of the main curve. When the scanning curve reaches the intersection with the main curve, the state of the soil will proceed along this main curve on further wetting. This behavior is especially clear in the wetting branch of the low-density soil shown in Figure 3, in which a decrease in suction starting from the as-compacted condition (w = 12%) to a suction of s = 10 kPa causes only a small increase in water content. A similar behaviour was observed on wetting by Delage & Suraj de Silva (1992), testing a compacted silt.

3 EXPERIMENTAL PROGRAM

3.1 Controlled-suction equipment

An improved controlled-suction triaxial cell similar to the equipment described in Romero et al. (1997), Romero (1999) and Barrera (2002) was used to perform the tests. Figure 4 shows a schematic layout of the cell and the experimental setup. Dimensions of the specimen are 38 mm in diameter and 76 mm in height. The deformation response was monitored with local axial (miniature LVDTs, with a measurement range of ±3 mm, adhered to the membrane) and radial (electro-optical laser system mounted on two diametrically opposite sides) transducers. Local vertical strains can be measured in the range from 10^{-5} to 10^{-1}, moreover at higher strains the measurements are performed by means of an external LVDT and including corrections due to cell deformability. Vertical profiles of specimen may be obtained by moving the laser sensor by means of an electrical motor. These profiles allow obtaining precise values of global volume changes of the specimen. The main characteristics of the measurement systems used in the tests are summarized in Table 2.

Suction was applied simultaneously via axis translation technique on both ends of the sample, maintaining a constant air pressure and modifying the water pressure. Both top and bottom platens have a combination of two porous discs: a peripheral annular coarse one connected to the air system and an internal disc with a high air-entry value ceramic (1.5 MPa) connected to the water system. This double drainage ensured a significant reduction of the equalization time.

Water content changes were registered measuring the water volume that crossed both HAEV discs by means of two double wall burettes with differential pressure transducers. The measured water volume changes were corrected taken into account the amount of air diffused through the ceramic discs and the leakage through the pipes. In this way, delicate effects concerning the inflow and outflow of water during loading could be successfully examined.

Two stepper motors operating air pressure regulators were used to continuously control the deviator and confining stresses. The stepper motors and the measurements of 14 sensors are managed by an automatic data acquisition and control system that allows applying general stress and suction paths and performing strain controlled tests.

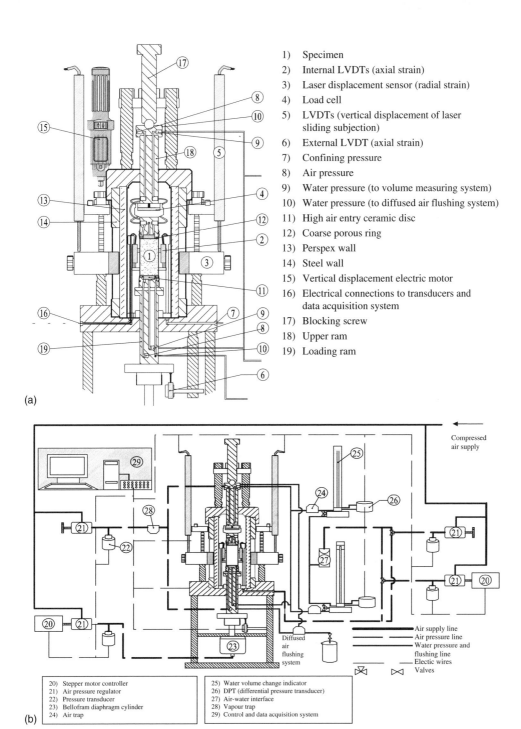

Figure 4. Experimental apparatus (a) scheme of triaxial cell, (b) layout of experimental set up for suction and stress controlled triaxial test: pressure control system and measuring system.

Table 2. Properties of the different transducers.

Measured variables (type)	Transducer	Operating range (FS)	A/D resolution (1 lsb)	Non-linearity +hysteresis (% FS, engineering units)
u_a, u_w, σ_r, σ_p^* (pressure)	diaphragm	2500 kPa	0.08 kPa	0.13%, 3 kPa
q (load)	strain gauge cell	9.8 kN (8.8 MPa)**	0.3 N (0.3 kPa)**	0.12%, 12 N, (11 kPa)**
ε_a internal (displacement)	LVDT	6 mm	0.2 μm	0.20%, 12 μm
ε_a external (displacement)	LVDT	15 mm	0.5 μm	0.20%, 60 μm
ε_r (displacement)	laser-based electro-optical	5 mm	2 μm	0.21%, 11 μm
Disp. Vert. laser (displacement)	LVDT	100 mm	3 μm	0.20%, 200 μm
V_w (differential pressure)	eddy current sensor	1.5 kPa (19 cm³)***	0.05 Pa (0.5 mm³)***	0.21%, 3 Pa (41 mm³)***

* u_a, u_w, σ_r, σ_p: pressure air pressure, water pressure, confining pressure and piston pressure, respectively.
** Vertical stress range, resolution and accuracy (sample diameter of 38 mm).
*** Sensitivity, range, resolution and non-linearity + hysteresis in burette.

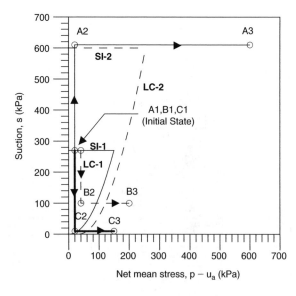

Figure 5. Stress path followed on the clayey silt. Yield curve evolutions.

3.2 *Stress paths followed*

Three isotropic compression tests (A, B, and C shown in Fig. 5) were performed at different suctions. Equalization stages A1-A2, B1-B2 and C1-C2, to bring the as-compacted suction to the different target suctions, are shown in Figure 5 (A1, B1, C1 represent the as-compacted states). Isotropic compression paths (A2-A3, B2-B3 and C2-C3) are also indicated.

Equalization stages were carried out to apply target suctions of 600, 100 and 10 kPa (A2, B2 y C2 respectively). Equalization was assumed to be achieved when water content and deformations became stable, or according to Sivakumar (1993) and Rampino et al. (1999), once water flow was lower than water content changes of 0.04% per day.

Equalization period was about 250 hours at a target suction of 10 kPa. All the equalization stages were performed under a constant mean net stress $(p - u_a) = 20$ kPa and a deviator stress $q = 10$ kPa. These low values were chosen to avoid collapse on wetting and to allow the detection of the yield stress $(p_0 - u_a)$ in the subsequent isotropic compression path under constant suction. Some small collapse (irreversible deformation) was observed when suction was maintained below 10 kPa in the oedometer test used to find the retention curve of the low-density soil. This fact was associated with the dragging of the 'loading-collapse' LC yield locus, as proposed by Alonso et al. (1990). The initial position of the LC-1 curve that corresponds to the as-compacted state is shown in Figure 5. As observed, the wetting paths under isotropic conditions evolved in the elastic domain, developing small reversible swelling strains. On the other hand, during the drying path under isotropic conditions to reach a suction of 600 kPa, the soil underwent irreversible shrinkage that was associated with the dragging of the 'suction increase' SI yield surface, as proposed by Alonso et al. (1990). The strain hardening induced by this drying process, which enlarged the elastic domain, was also reflected by the new position of the LC-2 curve indicated in Figure 5. The initial (SI-1) and final positions (SI-2) of the SI yield loci are shown in Figure 5.

The increase of mean net stress $(p - u_a)$ was applied at a stress rate of 1.8 kPa/hr, under a constant deviator stress $q = 10$ kPa. This stress rate was considered adequate to avoid water pressure build-up. This condition was verified when measuring negligible compression strains after maintaining a constant confining stress for at least 24 hours at the end of the compression ramp. The maximum mean net stress of each test was chosen to determine the yield stress at different suctions and to display enough post-yield response due to the dragging of the LC yield locus.

4 EXPERIMENTAL RESULTS

Figure 6 displays the time evolution of different volumetric variables during the suction equalization stages. The following volumetric variables were selected: void ratio e, water ratio e_w and degree of saturation $(Sr = e/e_w)$. Water inflow and some swelling were observed when suctions of 10 kPa and 100 kPa were applied, and water outflow and shrinkage were measured when suction was increased to 600 kPa. The small changes detected in the equalization stage at $s = 100$ kPa indicated that the initial as-compaction suction was close to this value. During the suction increase path it was easier to expel water than induce shrinkage deformation on soil skeleton, and e_w/e reduces. It was also admitted that when suction increased over the SI-1 yield locus (refer to Fig. 5), which bounds the transition between elastic and virgin states, both simultaneous irreversible strains and irrecoverable water ratios developed affecting in a poroplastic way soil behaviour (Vaunat et al. 2000).

Figure 7 shows the changes of the different volumetric variables undergone by the soil during the different compression paths. As observed, the evolution of variable e displayed clear pre- and post-yield zones. Yield stresses increase at higher suctions in accordance with the elastoplastic model of Alonso et al. (1990). Post-yield response on variable e was associated with the dragging of the loading-collapse LC yield curve that was sketched in Figure 5. The post-yield compressibility decreased at higher suctions, also in accordance with the model. A common yield stress in the $e:\ln(p - u_a)$, $e_w:\ln(p - u_a)$ and $e/e_w:\ln(p - u_a)$ planes was identified for all the volumetric variables along the compression paths at suctions of 100 and 600 kPa. In these paths, the evolution of the degree of saturation displayed an increasing trend on loading in the post-yield range, as a consequence of the higher efficiency of the loading mechanism in deforming soil skeleton (macro-pore volume reduction) than expelling water (emptying of macro-pores). No significant degree of saturation changes were detected in the pre-yield range of these paths. These experimental tendencies were similar to those reported by Rampino et al. (1999, 2000).

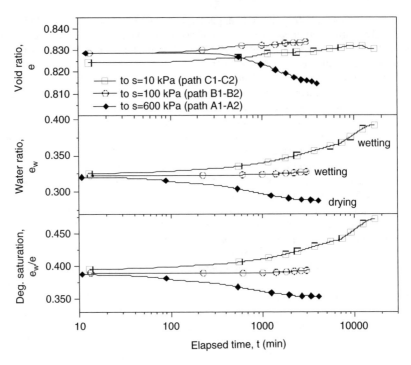

Figure 6. Evolution of void ratio, water ratio and degree of saturation during the suction equalization stages.

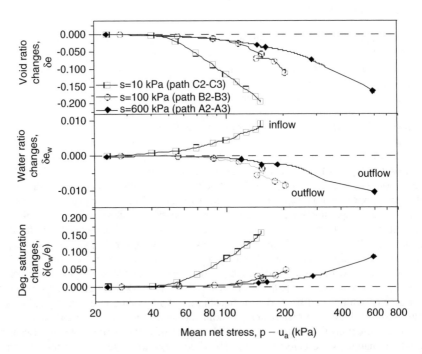

Figure 7. Changes in void ratio, water ratio and degree of saturation during the isotropic compression paths.

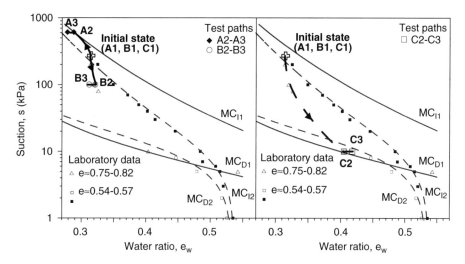

Figure 8. Changes in void ratio, water ratio and degree of saturation during the isotropic compression paths.

The same figure shows a clear tendency of water ratio increase in the loading path at $s = 10\,\text{kPa}$. In this case, the important increase of degree of saturation was associated with two mechanisms: soil skeleton deformation due to the higher post-yield compressibility (macropore volume reduction) and flooding of macropores. The second mechanism was a consequence of the important macropore volume and size reduction undergone by the material on loading. This new pore network was more eager for retaining water, due to the higher air-occlusion value of the wetting branch of the retention curve induced by the decrease of the void ratio (refer to Fig. 7). Sivakumar (1993) presented test results that also displayed water inlet during a ramped compression.

The hydraulic response can be plotted in an $e_w:\ln(s)$ plane. In Figure 8, the 'main wetting and drying curves' for the as-compaction state ($e_1 = 0.75 - 0.82$) are indicated by MC_{D1} (Main Curve for suction Decrease) and MC_{I1} (Main Curve for suction Increase). These bounding wetting and drying curves at a constant void ratio enclose the scanning region and separate attainable states (in-side this region) from unattainable states. As pointed out previously, when the soil undergoes a reduction in the macropore volume, the shape of these main curves changes (MC_{D1} to MC_{D2} and MC_{I1} to MC_{I2}, as shown in Fig. 8), increasing their air-entry value on drying and their air-occlusion value on wetting, and decreasing the water ratio at saturation.

In the first test at high suction (A1-A2-A3 in Fig. 8), the soil shows an important macropore volume reduction on suction equalization (shrinkage A1-A2) and the subsequent isotropic compression path A2-A3. The volumetric strains undergone by the soil, induced the movement of the main drying curve from MC_{I1} to MC_{I2}. During the suction increase stage, the state of the sample moved initially on a 'scanning curve' until the MC_{I1} curve was reached. Afterwards, the state of the soil remained on this bounding curve and it followed its movement. On loading, due to the constraint that the zone on the right side of MC_I is unattainable, the soil state is pushed (path A2-A3) by the movement of the main drying curve, causing a small decrease in water ratio at constant suction.

In the second test at medium suction (B1-B2-B3 in Fig. 8), the changes in the main drying and wetting curves were negligible during the suction equalization stage (B1-B2), because the volumetric strains undergone by the soil were small. The state of the soil during this wetting path remained inside the attainable zone, following a 'scanning curve' with a small increase in water ratio. Vaunat et al. (2000) assumed a reversible response within this scanning zone. During the subsequent isotropic compression path B2-B3, the void ratio decreased from $e = 0.83$ to 0.72, inducing a slight movement of the main curves. However, the state of the soil still remained inside the scanning zone between both main curves. Within this zone, Romero & Vaunat (2000) and Vaunat et al. (2000) assumed a reversible response of water ratio changes induced by loading/unloading paths. The elastic hydraulic stiffness against net stress changes proposed by these authors predicted a water ratio decrease on loading, which was the same response observed in the loading path B2-B3.

In the third test performed at very low suction (C1-C2-C3 in Fig. 8), the changes in MC_{D1} and MC_{I1} were also negligible during the suction equalization stage (C1-C2), because the swelling strains undergone by the soil were small. As in the previous test, the soil initially followed an elastic path over a 'scanning curve' ending at a final state C2, which was probably near the main wetting curve MC_{D1}. During the compression path C2-C3, important plastic volumetric strains were developed, which induced the movement of the main wetting curves from MC_{D1} to MC_{D2}. In this case, it is particularly important the increase in the air-occlusion value of the soil due to the decrease in the pore diameter caused by the compression process. As observed, the new position MC_{D2} intersects the initial MC_{D1} at $s < 10\,kPa$. It was assumed that the state of the soil remained on this bounding wetting curve and it followed its movement. On loading, due to the constraint that the zone on the left side of MC_D is unattainable, the soil state is pushed (path C2-C3) by the movement of the main wetting curve, causing an increase in water ratio at constant suction.

5 CONCLUSIONS

Fabric descriptions by means of MIP and ESEM observations were performed on silty clay samples compacted at two different dry densities, which correspond to different loading states, in order to observe the fabric evolution during the loading processes. In addition, retention curves for compacted states at both contrasting structures were obtained to show the main effects induced by this compression process.

Both qualitative description by ESEM tests and quantitative description by MIP tests show reasonably consistency in the fabric description. The pore size distribution is clearly bi-modal. The larger voids would correspond to the inter-aggregate pores and compaction stress increase affects mainly the pore structure of these inter-aggregate pores. On the other hand, intra-aggregate pores appear largely independent of compaction effort.

Additional information on the soil fabric can be inferred from the examination of retention curves. It can be observed that the water retained at high suctions is independent of the dry density. It seems plausible that the water retained at those high suctions belongs to the intra-aggregates pores where the total porosity plays no relevant role.

The comparison between the retention curves illustrates its dependence on void ratio, which affects the water storage capacity of the soil at saturation, the air-entry value on drying and the air-occlusion value on wetting.

A series of three isotropic compression paths at different suctions were performed in a fully-instrumented triaxial cell to study the coupled hydro-mechanical response of the silty clay, which was statically compacted at a very low dry density. This low value was selected to induce an appreciable change of the void ratio and the water retention properties of the soil on loading.

The experimental technique, involving suction equalization and ramped compression stages, was described in detail. The local instrumentation of axial and radial strains installed in the triaxial cell, as well as the continuous recording of the soil water volume changes with automatic burettes, allowed the monitoring of the coupled hydro-mechanical response and the successful examination of delicate effects concerning the inflow and outflow of water during loading.

The experimental results showed the important role played by the mechanical path in modifying the water retention properties of the soil. Three equalization stages were selected at suctions 10, 100 and 600 kPa, which corresponded to a zone in the retention curve plot close to the bounding 'main wetting curve', a zone in the 'scanning domain' between both main curves, and a zone close to the bounding 'main drying curve'. On loading at a suction of 100 and 600 kPa, the sample expelled water, whereas at a suction of 10 kPa there was a clear tendency to absorb water due to the formation of a denser structure with a higher air-occlusion value. This fact is explained in terms of the movement and the change of shape on loading of the bounding drying and wetting curves, which depend on the void ratio and delimit the domain of attainable states.

Mechanical results at different suctions, such as the post-yield compressibility and the yield stress of the loading paths, were interpreted within the framework of the elastoplastic model of Alonso et al. (1990). Water content changes observed on isotropic loading were interpreted within the framework of bounding retention curves proposed by Vaunat et al. (2000), which separate a domain of attainable states from unattainable states in the water content-suction plane.

ACKNOWLEDGEMENTS

The first author acknowledges the financial support provided by Universidad Autónoma de Campeche (México) and PROMEP grant from SEP (México). The support of the Spanish Ministry of Science and Technology through research grant BTE2001-2227 is also acknowledged.

REFERENCES

Alonso, E.E., Gens, A. & Josa, A. 1990. A constitutive model for partially saturated soils. *Géotechnique* 40 (3): 405–430.
Barbour, S.L. 1998. Nineteenth Canadian Geotechnical Colloquium: The soil-water characteristic curve – A historical perspective. *Can Geotech J* 35: 873–894.
Barrera, M. 2002. Estudio experimental del comportamiento hidro-mecánico de suelos colapsables (in Spanish). Ph.D. thesis, Universitat Politècnica de Catalunya, Barcelona, Spain.
Chen, Z.-H., Fredlund, D.G. & Gan, J.K.-M. 1999. Overall volume change, water volume change, and yield associated with an unsaturated compacted loess. *Can Geotech J* 36: 321–329.
Childs, E.C. 1940. The use of soil moisture characteristics in soil studies. *Journal of Soil Science* 50: 239–252.
Delage, P., Audigier, M., Cui, Y.-J. & Howat, M.D. 1996. Microstructure of a compacted silt. *Can Geotech J* 33: 150–158.
Delage, P. & Suraj de Silva, G.P.R. 1992. Negative pore pressure and compacted soils. In: E. Ovando, G. Auvinet, W. Paniagua & J. Díaz (eds), *Raul J. Marsal* Vol.: 225–232. México: Sociedad Mexicana de Mecánica de Suelos.
Karube, D. & Kawai, K. 2001. The role of pore water in the mechanical behaviour of unsaturated soils. *Geotechnical and Geological Engineering* 19: 211–241.
Kawai, K., Nagareta, H., Hagiwara, M. & Iizuka, A. 2003. Suction changes of compacted soils during static compaction test. In: D. Karube, A. Iizuka, S. Kato, K. Kawai & K. Tateyama (eds), *Unsaturated soils. Geotechnical and geoenvironmental issues; Proceedings of the 2nd Asian conference on unsaturated soils, Osaka*: 429–434. Japan.
Rampino, C., Mancuso, C. & Vinale, F. 1999. Laboratory testing on unsaturated soil: equipment, procedures and first experimental results. *Can Geotech J* 36: 1–12.
Rampino, C., Mancuso, C. & Vinale, F. 2000. Experimental behaviour and modelling of an unsaturated compacted soil. *Can Geotech J* 37: 748–763.
Ridley, A.M. & Burland, J.B. 1993. A new instrument for the measurement of soil moisture suction. *Géotechnique* 43(2): 321–324.
Romero, E., Faccio, J.A., Lloret, A., Gens, A. & Alonso, E.E. 1997. A new suction and temperature controlled triaxial apparatus. *Proc 14th Int Conf on Soil Mechanics and Foundation Engineering, Hamburg*: 185–188. Rotterdam: Balkema.
Romero, E. 1999. Characterization and thermo-hydro-mechanical behaviour of unsaturated Boom clay: an experimental study. Ph.D. thesis, Universitat Politècnica de Catalunya, Barcelona, Spain.
Romero, E., Gens, A. & Lloret, A. 1999. Water permeability, water retention and microstructure of unsaturated Boom clay. *Engineering Geology* 54: 117–127.
Romero, E. & Vaunat, J. 2000. Retention curves of deformable clays. In: A. Tarantino & C. Mancuso (eds), *International Workshop On Unsaturated Soils: Experimental Evidence and Theoretical Approaches in Unsaturated Soils, Trento, Italy*: 91–106. Rotterdam: A.A. Balkema.
Sivakumar, V. 1993. A critical state framework for unsaturated soil. Ph.D. thesis, University of Sheffield, Sheffield, U.K.

Tombolato, S., Tarantino, A. & Mongiovì, L. 2003. Suction induced by static compaction. *International conference from experimental evidence towards numerical modelling of unsaturated soils; Weimar, Germany*: 1–10.

Vanapalli, S.K., Frendlund, D.G. & Pufahl, D.E. 1999. The influence of soil structure and stress history on the soil-water characteristic of a compacted till. *Géotechnique* 49(2):143–159.

Vaunat, J., Romero, E. & Jommi, C. 2000. An elastoplastic hydro-mechanical model for unsaturated soils. In: A. Tarantino & C. Mancuso (eds), *International Workshop On Unsaturated Soils: Experimental Evidence and Theoretical Approaches in Unsaturated Soils, Trento, Italy*: 121–138. Rotterdam: A.A. Balkema.

An experimental study on a partially saturated pyroclastic soil: the Pozzolana Nera from Roma

E. Cattoni, M. Cecconi & V. Pane
Università di Perugia, Italy

ABSTRACT: The paper presents the results of an experimental investigation on the mechanical behaviour of partially saturated *Pozzolana Nera* (Roma, Italy). Such natural deposits, which have been extensively studied in saturated conditions, are characterised by a marked heterogeneity in terms of grading, nature of grains, and inter-particle bonds. In situ, deposits of pozzolana are generally found in unsaturated conditions; this has an important practical relevance in the evaluation of the stability conditions of natural slopes and cuts. The experimental investigation consisted mainly of isotropic and triaxial compression tests on reconstituted samples, at increasing values of mean net pressures in the range 50–400 kPa, at different constant levels of suction (20–75 kPa). Volume pressure plate extractor tests were also conducted to obtain the water retention curve. In addition some *wetting* tests were carried out by means of conventional oedometer cells. The testing programme made it possible to examine the hydraulic and mechanical properties of unsaturated *Pozzolana Nera* at different values of the degree of saturation and to compare the mechanical behaviour of the same material in saturated conditions. Special attention was focused on defining the failure envelope at relatively low confining stress and assessing the influence of the degree of saturation on the failure conditions of the material.

1 INTRODUCTION

The inspiration for this work stems from the wide set of experimental results so far gathered from a laboratory investigation on the mechanical behaviour of *Pozzolana Nera*, a bonded coarse-grained material, which originated from the explosive activity of Colli Albani volcanic complex (Upper-Middle Pleistocene), about 25 km south-east of Roma. The experimental study on saturated natural *Pozzolana Nera* started about some years ago; the research was driven by need of investigating the stability of sub-vertical cuts and underground cavities, very frequent in these deposits. The main outcomes of the study are described in detail in Cecconi & Viggiani (1998), Cecconi (1999), Cecconi & Viggiani (2000, 2001).

However, since the material *in situ* is only partially saturated ($S_r = 0.4 \div 0.5$), further research and experimental work on the hydraulic and mechanical properties of partially saturated material seemed to be compulsory for both basic research and engineering applications. In this paper some of the main experimental results obtained on reconstituted partially saturated *Pozzolana Nera* from oedometer and triaxial compression tests, as well as data obtained from pressure plate volume extractor (Cattoni, 2003), are presented and critically discussed within the frame-work of recent studies on this topic.

2 MATERIAL TESTED AND EXPERIMENTAL PROGRAMME

The *Pozzolana Nera* is a coarse grained pyroclastic weak rock originating from pyroclastic flows chaotically deposited at high temperature in response to gravity and then cooled.

Natural deposits of pozzolanas contain crystals, glass shards and pumices, lithic fragments in highly variable proportion. The microstructure consists of sub-angular grains of very variable size with a rough surface; intrinsic inter-particle bonds, probably due to the original material continuity at time of deposition, are made of the same constituents of grains and aggregates. Therefore, bond deterioration and grain crushing upon loading can be considered as the same response to mechanical loading.

At this regard, it is worth to mention some main features of the mechanical behaviour of the natural material, in saturated condition. In isotropic and 1D compression, gradual yield results from grain crushing and/or progressive breakage of inter-particle bonds, since bonds and grains are made of the same material. The initial porosity is the main factor controlling compressibility. Upon shearing, the mechanical response of the material changes from brittle and dilatant to ductile and contractant with increasing confining stress; for both brittle and ductile behaviour, failure is associated with the formation of noticeable shear surfaces separating the sample into rigid bodies. The experimental data indicate that peak strength does not correspond to the maximum rate of dilation and, in addition, the condition of null dilatancy, which in classical critical state models defines the friction of the material, is attained at two different stress ratios (Cecconi & Viggiani, 2001). Based on the strong physical assumption that progressive bond deterioration and grain crushing induce a reduction of the friction angle of the material, a constitutive model has recently been developed in the framework of classical strain-hardening elasto-plasticity (Cecconi et al., 2002, 2003).

The laboratory testing on partially saturated *Pozzolana Nera*, herein described, consisted of volume extractor tests, oedometer/wetting tests and triaxial compression tests at constant suction, performed mainly on reconstituted samples; a few tests have been also performed on natural samples.

Soil-water retention curves (*SWCC*) were evaluated by using the standard volumetric pressure plate device, manufactured by SOIL MOISTURE EQUIPMENT INC. (Santa Barbara, California). Initially saturated samples were prepared by moist tamping in a 112 mm dia. and 10 mm high aluminium mould. The material was preliminarily submerged in distilled water for 24 hours, undergoing several cycles of vacuum. The volumetric pressure plate extractor is provided with a 200 kPa a.e.v ceramic disk ($k = 3 \cdot 10^{-7}$ cm/s) located at its base. Suction increments/decrements (Δs) are applied by increasing/decreasing the air pressure in the cylinder, while the pore water pressure is kept constant at atmospheric pressure. The air pressure is regulated by two pressure converters installed in series and measured by means of a digital precision manometer with an accuracy of ± 0.07 kPa and a full range of 10 kPa. For each suction increment (or decrement) the volume water removed (*drying* path) or adsorbed (*wetting* path) can be measured through a glass burette, after equalization is reached. Typically the equalization stages required approximately 2–3 days for each suction increment.

Oedometer tests were carried out in conventional cells varying from 35.7 up to 71.4 mm in diameter. The material was first oven-dried, mixed with de-aired water and compacted in thin layers, directly into the mould. Changes in initial water content were minimised by covering the cell with a cellophane film. Wetting tests were performed at constant vertical stress, after primary settlements had occurred ($dh/dt \leq 1.7 \cdot 10^{-5}$ mm/min).

Finally, triaxial compression tests were carried out by means of a servo-controlled automated system for unsaturated soils equipped with a Bishop & Wesley triaxial cell, manufactured by MEGARIS (Caserta, Italy). The prototype of this apparatus has been first set up by Nicotera (1998) and it is fully described in Nicotera et al., 1999 and Aversa & Nicotera (2002).

The system has been designed to test 50 mm in diameter, 100 mm high cylindrical samples. The cell fluid is air; hence the outer Perspex cylinder is enclosed by a steel cylindrical shield. Suction is controlled by applying, controlling and measuring independently positive values of pore air and pore water pressures (*axis translation* technique). Axial strains were measured by means of external LVDT transducers, while volume strains were inferred from measured radial strains. The cell base is also provided to mount local miniaturised LVDT's aimed to measure locally both axial and radial strains (Cuccovillo & Coop, 1998). The system used to determine radial strains is made of an inner water-filled aluminium cylinder surrounding the sample; any variations of water level is related to the deformation of the sample. A glass burette, filled with water and kept at the same cell pressure,

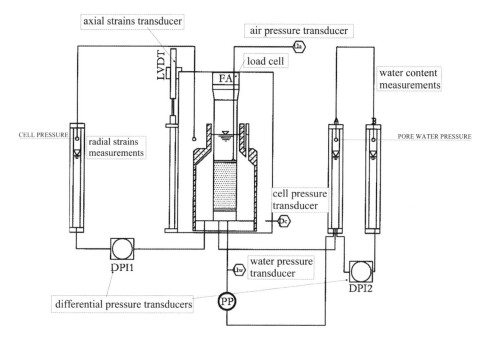

Figure 1. Measurement system in the triaxial equipment (adapted from Nicotera et al., 1999).

is used to fix a reference. Then a differential pressure transducer (accuracy of $6.0 \cdot 10^{-3}$ kPa) is used to measure the difference in pressure between the water contained in the inner cell and the reference burette (Figure 1). Upon shearing, the average radial strain is related to the water level changes in the inner cell (ΔI_r) as follows (Aversa & Nicotera, 2002):

$$\varepsilon_r = -\frac{1}{2}\frac{A_b - A_{s0}}{V_{s0}}\Delta I_r \qquad (1)$$

where A_b and A_{s0} represent respectively the cross section of the inner cell and the initial sample cross section; V_{s0} is the initial volume sample. The equipment adopted to measure water content changes (ΔI_w) consisted of two burettes, one connected to the drainage circuit at the base of the pedestal, below the h.a.e.v. porous stone, the other used as a reference level (see Figure 1). Water filling the burettes is kept at the same water pressure u_w. Also in this case, a differential pressure transducer is used to measure the level difference ΔI_w. Changes in volumetric water content ($\Delta\theta_w$) are then calculated as follows:

$$\Delta\theta_w = -\Delta I_w \frac{A_{bw}}{V_{s0}} \qquad (2)$$

where A_{bw} represents the burette cross section (5 mm dia). A change of water level of 0.1 mm corresponds to a volumetric water content change of $10^{-3}\%$.

Triaxial compression tests were carried out following three different stages: (i) increase of suction at constant cell pressure, (ii) isotropic compression at constant suction, (iii) strain-controlled drained shearing at constant cell pressure and suction. Isotropic compression was performed in continuous loading at a rate of 10 kPa/h, while shearing was carried out at displacement rates of about 0.00417 mm/min ($\cong 0.25\%$/hour).

Table 1 summarises the initial values of water content (w_0), voids ratio (e_0), degree of saturation (S_r) of all samples, while the initial grain size distribution is shown in Figure 2. In this re-constituted state, the material can be classified as a silty sand ($U = 3 \div 18$).

Table 1. Laboratory tests: initial physical properties of samples. EV: volume extractor tests; OED: oedometer tests; TX: triaxial compression tests.

Type	#Test	w_0	e_0	S_{r0}	θ_{w0}	s (kPa)	$p_{net}^{(1)}$ (kPa)	$\sigma_{v,wetting}$ (kPa)	$\Delta\varepsilon_{a,wetting}$ (%)
EV	EV03REC	0.34	0.90	1.00	0.47	//	//	//	//
	EV04REC	0.27	0.73	1.00	0.43	//	//	//	//
	EV05REC	0.31	0.83	1.00	0.45	//	//	//	//
OED	EDO01REC	0.14	1.25	0.30	0.17	//	//	100[2]	0.094
	EDO02REC	0.13	1.13	0.32	0.17	//	//	400[2]	0.215
	EDO03REC	0.14	1.03	0.36	0.18	//	//	1600[2]	0.149
	EDO04REC	0.13	1.01	0.35	0.18	//	//	800[2]	0.400
	EDO05REC	0.10	1.15	0.23	0.13	//	//	3200[2]	1.079
	EDO06EC	0.13	1.60	0.21	0.13	//	//	400[3]	1.837
	EDO07REC	0.13	1.62	0.21	0.13	//	//	400[4]	1.022
	EDO08REC	0.14	1.45	0.25	0.15	//	//	400[3]	1.801
	EDO09REC	0.13	1.50	0.24	0.14	//	//	400[4]	−0.025
	EDO10REC	0.13	1.60	0.21	0.13	//	//	50[5]	0.094
TX	TX03REC	0.23	0.63	1.00	0.39	20	200	//	//
	TX06REC	0.25	0.68	1.00	0.40	20	400	//	//
	TX07REC	0.26	0.70	1.00	0.41	20	100	//	//
	TX12REC	0.24	0.65	1.00	0.40	20	50	//	//
	TX10REC	0.25	0.68	1.00	0.41	75	100	//	//
	TX11REC	0.24	0.63	1.00	0.39	75	50	//	//
	TX13REC	0.26	0.70	1.00	0.41	75	200	//	//
	TX14REC	0.22	0.60	1.00	0.38	//	200	//	//
	TX15REC	0.26	0.69	1.00	0.41	75	100[6]	//	//
	TXMC	0.39	1.05	1.00	0.51	//	200	//	//

[1] At the end of isotropic compression; [2] wetting upon loading; [3] wetting after unl-rel cycle; [4] wetting before unl-rel cycle; [5] after unloading; [6] suction applied after isotropic compression.

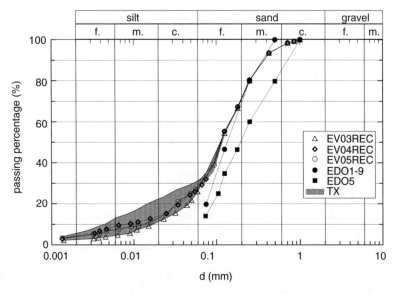

Figure 2. Grain size distribution of reconstituted samples of *Pozzolana Nera*.

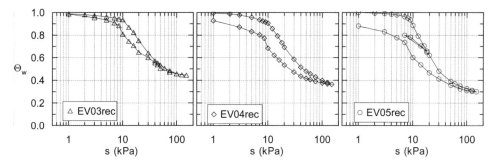

Figure 3. Soil water retention curves for reconstituted *Pozzolana Nera*.

3 HYDRAULIC PROPERTIES

3.1 Soil water retention curves

The volumetric pressure plate device allowed to assess the soil retention curve of partially saturated *Pozzolana Nera* for values of suction smaller than the *h.a.e.v.* of the ceramic disk (200 kPa). Initial values of physical properties of tested samples are reported in Table 1.

Figure 3 shows the experimental *SWCC*s obtained from three tests in terms of normalised volumetric water content ($\Theta_w = \theta_w/\theta_{ws}$) as a function of suction. The air entry value (*a.e.v.*) of the soil can be estimated from the well defined kink of the curve, by using the procedure proposed by Fredlund & Xing (1994). All *SWCC*s are characterised by a relatively narrow hysteretic domain; the *a.e.v.* ($\cong 8.5$–9 kPa) is approximately the same for all the samples, thus indicating that the effect of initial grain size distribution overwhelms the difference in initial voids ratio ($e_0 = 0.73 \div 0.90$, see Table 1). In fact, the *a.e.v* is directly controlled by the maximum grain diameter d_{max} which is almost identical for the three samples, while small differences in fine content result in different wetting-drying paths.

At present, several equations for the soil-water retention curves are available in the literature. Among these, the equation proposed by Fredlund & Xing (1994) and the one proposed by Van Genuchten (1980) have been chosen to fit the experimental data, by using a non-linear curve fitting algorithm (MATLAB 6.1).

The equation proposed by Fredlund & Xing (1994) is:

$$\Theta_w = \frac{1}{\left[\ln\left(e_N + \left(\frac{s}{a}\right)^n\right)\right]^m} \qquad (3)$$

with $e_N = \exp(1)$. In equation (3) parameters *a*, *m* and *n* are defined as follows: parameter *m* depends on the shape of the *SWCC* close to the inflection point; *a* corresponds to the inflection point of the curve, and for small values of *m*, is equal to the *a.e.v.*; parameter *n* depends on the shape of the pore size distribution and increases with increasing the uniformity of the pore size distribution.

On the other hand the equation proposed by Van Genuchten is:

$$\Theta = \frac{1}{\left[(1+(a's)^{n'})\right]^{m'}} \quad \text{with} \quad \Theta = \frac{\theta_w - \theta_{wr}}{\theta_{ws} - \theta_{wr}} \qquad (4)$$

The quantity θ_{ws} is the volumetric water content in saturated conditions, while θ_{wr} is the residual volumetric water content which has been estimated as the value of θ_w at $s = 150$ kPa.

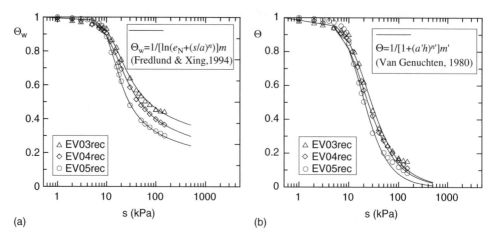

Figure 4. Experimental data and curve fittings; (*a*) Fredlund & Xing model; (*b*) Van Genuchten model.

Table 2. Curve fitting parameters and measured a.e.v.s.

Test #	a.e.v. (kPa)	Fredlund & Xing (1994)			Van Genuchten (1980)		
		a (kPa)	m	n	a' (m^{-1})	n'	θ_{wr}
EV03REC	9.0	11.90	0.43	2.81	0.6	2.030	0.16
EV04REC	9.0	11.40	0.54	2.51	0.7	2.030	0.12
EV05REC	8.5	10.85	0.54	3.65	0.7	2.310	0.11

Similarly, three parameters a', m', and n' compare in Equation 4 but only two of them are independent since $m' = 1 - 1/n'$ (Mualem, 1976). Parameters a' and n' have the same physical meanings of a and n in Equation (3).

In Figures 4a and 4b the experimental drying paths are fitted using eqs (3) and (4). It is noted that the equation proposed by Fredlund & Xing fits nicely the experimental data in the whole range of suctions, while Equation 4 seems to underestimate the experimental data for suctions larger than 60 kPa. The two sets of parameters entering Equations (3) and (4) are reported in Table 2.

3.2 *Conductivity function*

Three different approaches have been pursued in order to evaluate the conductivity function of *Pozzolana Nera*:

(a) analytical methods based on the solution of Richards' equation

$$\frac{d\theta_w}{dt} = \frac{\partial k_w}{\partial z} + \frac{\partial}{\partial z}(D_w \frac{\partial \theta_w}{\partial z}) \tag{5}$$

for water flow in unsaturated soils (Gardner, 1956; Rijtema, 1959);
(b) semi-empirical methods which make use of the measured soil water characteristic curve (Kunze et al., 1968; Green & Corey, 1971; Nielsen et al., 1972);
(c) statistical methods based on the pore size distribution (Burdine, 1953; Mualem, 1976; Van Genuchten, 1980; Fredlund & Xing, 1994).

The methods belonging to the first class allow to estimate the conductivity function from the comparison between the experimental data – namely θ_w vs. time – obtained during the transient equalisation stage (pressure plate outflow data) and the theoretical solution derived by the integration of Richards differential equation, at given initial and boundary conditions. In these methods it is assumed that, for small increments of suction, $k_w(\theta_w)$ and the diffusivity coefficient $D_w(\theta_w)$ may be both taken constant. Therefore, the equation governing the one dimensional water flow induced by a change in suction, becomes formally identical to the equation of 1D consolidation (Terzaghi, 1923). In particular, the conductivity function of unsaturated *Pozzolana Nera* has been determined by using the methods proposed by Gardner (1956) and Rijtema (1959); these differ in that the latter takes into account the impedance of the *h.a.e.v.* ceramic disk.

In the second approach the experimental water retention curve is physically linked to a random pore size distribution. The conductivity function is given by a series obtained from the probability function of interconnections between water-filled pores of varying sizes:

$$k_w(\theta_w)_i = \frac{k_s}{k_{sc}} A_d \sum_{j=1}^{m} \{(2j+1-2i)s_j^{-2}\} \qquad (6)$$

where:

A_d: adjusting constant depending on the surface tension of water, the absolute water viscosity and the water density. It is assumed $A_d = 1$ (m·s^{-1}·kPa^2);
k_{sc}: computed saturated coefficient of permeability from Equation (6) when $k_w(\theta_w)_i = k_s$;
k_s: measured saturated coefficient of permeability ($k_s = 10^{-7}$ m/s).

The computational technique firstly prescribes a partition of the experimental *SWCC* into m intervals; each value of $k_w(\theta_w)_i$ represents the calculated water coefficient permeability for a specified volumetric water content $(\theta_w)_i$, corresponding to the mid-point of the *ith*-interval ($i=1$, 2, ..., m) where it is assumed to be constant.

Finally, the third approach for the prediction of $k_w(\theta_w)$ – see point c) above – involves the methods based on a reasonable approximate evaluation of the hydraulic conductivity of a pore domain with varying shape (*e.g.* Mualem, 1976). The conductivity function $k_w(\theta_w)$ is derived in a integral form and can be reduced to a closed form when a θ_w–s dependence is given by analytical expressions, such as the one proposed by Van Genuchten (1980):

$$k_w(h) = k_s \frac{\left\{1-(a'h)^{n'-1}\left[1+(a'h)^{n'}\right]^{-m'}\right\}^2}{\left[1+(a'h)^{n'}\right]^{m'/2}} \qquad (7)$$

In Equation (7), h is the suction potential ($h = s/\gamma_w$) and a', m', n' are the same parameters defined in Equation (4).

All three different approaches have been adopted in the evaluation of the conductivity function. In Figures 5a and 5b the values of $k_w(s)$ calculated through the analytical solutions of Gardner (1956) and Rijtema (1959) are compared with the values predicted by Equations (6) and (7). Such comparison shows that:

- the saturated coefficient of permeability k_s obtained with the analytical solution of Rijtema (1959) is in fair agreement with the measured value;
- k_w decreases of about 4 orders of magnitude in the range of suction 0–100 kPa ($S_r = 1 \div 0.3$), independently of the adopted method;
- the values of the conductivity coefficient obtained with the solution of Rijtema (1959) are in good agreement with those calculated with the method proposed by Green & Corey (1971), but the data are quite scattered at small suctions;
- the k_w values derived from the solution proposed by Gardner (1956) are less accurate, so indicating the important effect of the impedance of the *h.a.e.v.* ceramic disk;
- at large values of suction, the values of k_w computed with Equation (7) are sensibly smaller than those determined with the other two methods.

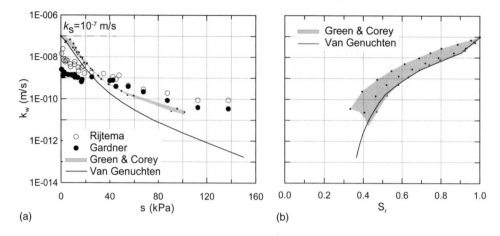

Figure 5. Conductivity function as a function of suction (*a*) and degree of saturation (*b*).

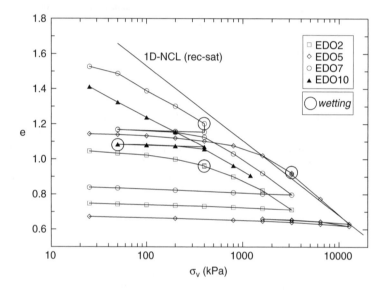

Figure 6. Oedometer tests with wetting paths at different vertical stress and voids ratios.

4 COMPRESSIBILITY

4.1 *One-dimensional compression with wetting paths*

Figure 6 shows the compressibility curves of four *wetting* tests performed in oedometer cells on partially saturated samples characterised by different initial voids ratios ($e_0 = 1.0 \div 1.6$), degree of saturations varying in a narrow range ($S_r = 0.2 \div 0.3$, see Table 1), and the same initial water content ($w_0 \cong 0.13$).

Upon loading, the samples were submerged in distilled water and saturated at constant values of vertical stress, σ_v. The majority of the samples exhibited a "collapsible" behaviour, whose amount depends on initial voids ratio, degree of saturation and stress level. In particular, for initial voids ratios in the range $e_0 = 1.01 \div 1.25$, the collapsible behaviour mainly depends on vertical stress; the larger is the vertical stress, the larger is the volume strain due sample saturation. No significant collapse-strains are observed for stress smaller than $\sigma_v = 3200$ kPa.

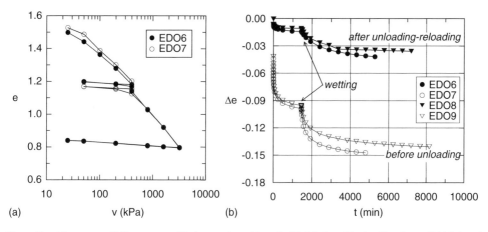

Figure 7. (*a*) compressibility curves; (*b*) changes in voids ratio (Δe) induced by loading ($\sigma_v = 400$ kPa) and following wetting.

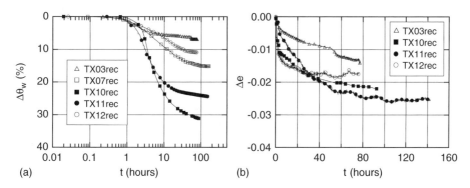

Figure 8. Equalisation stages. (*a*) volumetric water content changes *vs.* time; (*b*) changes in voids ratio (Δe) *vs.* time.

On the other hand, looser samples ($e_0 = 1.45 \div 1.60$), which are also characterised by smaller degrees of saturation, show a collapsible behaviour at lower stress levels ($\sigma_v = 400$ kPa). Therefore, the effect of the combination of initial voids ratio and degree of saturation seems to be larger than the one produced by the vertical stress (tests EDO5 and EDO7 in Figure 6 and Table 1).

The influence of unloading paths before wetting is shown in Figures 7*a* and *b*: samples saturated upon loading exhibited axial strains which are larger than those shown by the sample which underwent an unloading-reloading cycle before wetting (tests EDO6 and EDO7, see Table 1). In this case, the effect of previous unloading paths seems to prevail on that exerted by initial porosity; this is noted from Figure 7*b*) which shows the Δe-time curves induced by saturation for two couples of tests linked by the same load history but different initial porosity. Other tests show that if wetting occurs along an unloading path, or a reloading path well before reaching normal compression line (1D-NCL), axial strains induced by wetting may also be negative (swelling, see test EDO10rec after unloading). This behaviour was already observed by Alonso & Oldecop on rockfill (2001).

4.2 *Equalisation stages and isotropic compression*

In triaxial tests, isotropic compression was carried out only after the equalisation stage due to an increase of suction at constant cell pressure had been completed. Generally, when a change of suction is applied to the sample, subsequent strains occur due to the equalisation of pore pressures.

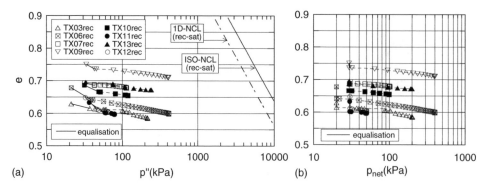

Figure 9. Equalisation stages and isotropic compression tests: (a) $e-p''$ (Bishop stress); b) $e-p_{net}$.

All samples of *Pozzolana Nera*, initially saturated and subjected to a mean net pressure $p_{net} \cong 30$ kPa, at applied suctions of 20 and 75 kPa exhibited significant positive volume strains. Figures 8 show the results obtained from selected equalisation stages performed throughout some tests. The results are plotted in terms of volumetric water content (Figure 8a) and changes in voids ratio (Figure 8b) against time. As the applied suction increases, the progressive decrease in both volumetric water content and voids ratio is more noticeable. Generally, equalisation stages took about one week to run out, independently of the imposed suction. It is noted that the reduction in voids ratio is rather large (corresponding to volume strains $\varepsilon_{v,max} \cong 1.75\%$), if compared to the overall strains built up during following isotropic compression and shear.

Experimental data obtained from isotropic compression are shown in Figures 9a and 9b as voids ratio versus the logarithm of effective stress p'' and mean net pressure p_{net}, where the effective stress p'' can be defined, following Bishop (1959), as:

$$p'' = p_{net} + S_r s \qquad (8)$$

The same figures show also the equalisation stages carried out at constant mean net pressure before isotropic compression. The variations in the degree of saturation occurring during these stages is clearly put in evidence from the oblique paths which are represented with continuous thick lines in Figure 9a). The one-dimensional and isotropic normal compression lines (1D and ISO-NCL) of saturated material are also reported in the same Figure. The behaviour of the material in isotropic compression is relatively stiff and no effects of applied suctions can be detected, may be due to the "overconsolidated" state of the material.

5 STRESS–STRAIN BEHAVIOUR AND STRENGTH

Figure 10a shows the experimental stress–strain curves obtained from three drained triaxial compression tests carried out at constant suction $s = 20$ kPa, in terms of deviator stress, q, versus deviatoric strain, $\varepsilon_s = 2/3 (\varepsilon_a - \varepsilon_r)$; the corresponding curves of volume strains, $\varepsilon_v = \varepsilon_a + 2\varepsilon_r$, versus ε_s are shown in Figure 10b. Similarly, the stress–strain curves obtained from other four triaxial tests at constant suction $s = 75$ kPa are plotted in Figures 10c and d.

For both series of data, it can be noted that both peak deviator stress and initial stiffness increase with the mean net pressure, p_{net}, as expected, while the deviatoric strain at peak is approximately the same for all tests. Data obtained from test TX07 are rather quivering, due to the malfunction of the load cell during the test. The soil behaviour is dilatant and the rate of dilation ($\delta\varepsilon_v/\delta\varepsilon_s$) slightly decreases with increasing p_{net}, as common for granular soils. The maximum rate of dilation is generally achieved before peak; this behaviour is different from the one observed for saturated natural *Pozzolana Nera*, for which maximum dilatancy was found to follow peak conditions (Cecconi & Viggiani, 2001). After peak, the deviator stress decreases towards an ultimate value (*eot* state).

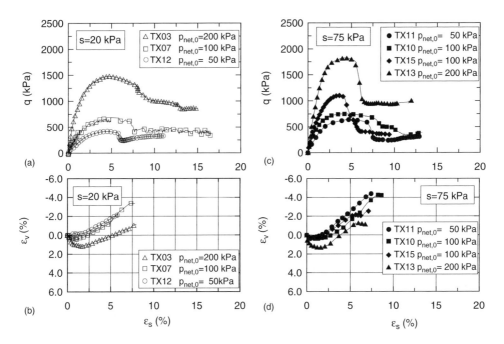

Figure 10. Triaxial compression tests at constant suctions $s = 20$ kPa and $s = 75$ kPa: (*a*) and (*c*) stress–strain curves; (*b*) and (*d*) volume strains *vs.* deviatoric strains.

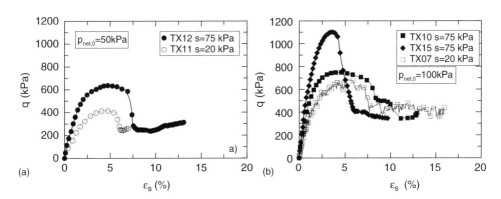

Figure 11. Stress–strain curves from triaxial compression tests at increasing suctions: (*a*) $p_{net,0} = 50$ kPa; curves; (*b*) $p_{net,0} = 100$ kPa.

Volume strains following peak conditions are not reliable and they have not been plotted in Figure 10, because of strain localisation occurring along shear bands.

In order to depict more clearly the effect of suction on the shear strength, the results of tests performed at suctions of 20 and 75 kPa, under a mean net pressure after isotropic compression $p_{net,0} = 50$ and 100 kPa, are reported in Figure 11. Data show visibly that the shear strength increases with suction and that this increase is quite about the 50% for tests at $p_{net} = 50$ kPa. In addition, although for tests TX10 and TX15 shearing was carried at the same value of suction $s = 75$ kPa, the maximum deviator stress observed during test TX15 is much larger than the one observed for test TX10. This behaviour can be ascribed to the different paths followed before shearing: the sample subjected to test TX15 had been firstly isotropically compressed, in saturated condition, and only after suction was increased (see Table 1).

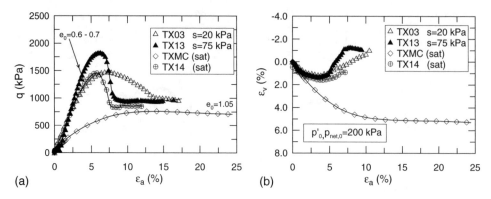

Figure 12. Triaxial compression tests on partially saturated and saturated samples: (*a*) stress–strain curves; (*b*) volume strains *vs.* axial strains.

The results obtained from two tests carried out on reconstituted saturated samples, isotropically consolidated at a value of mean effective stress $p' = 200$ kPa (tests TXMC and TX14) are shown in Figure 12 for comparison, in terms of deviator stress vs. axial strain ε_a (Figure 12a) and volume strains vs. axial strains (Figure 12b). From these plots it can be noted that: (*i*) the peak shear strength and the initial tangent stiffness both increase with increasing suction; (*ii*) no sensible effects of suction or degree of saturation can be detected on the ultimate shear strength; (*iii*) the end-of-test condition does not depend on initial voids ratio, as expected.

For all samples showing dilatant behaviour, failure was associated with very well defined shear surfaces separating the samples in two distinct bodies (see Figure 13). Strain-localization may have occurred quite immediately after peak; therefore the stress–strain data at larger strains than those corresponding to peak conditions must be regarded with very caution. Data corresponding to values of shear strains larger than about 7% have been intentionally omitted in Figures 10 and 12.

The shear strength of the material was examined in terms of stress parameters s, q, p_{net}. At peak conditions, the experimental results have been interpreted with the Mohr–Coulomb failure criterion, extended to unsaturated soils by Fredlund & Rahardjo (1993). The failure surface envelope is defined by the following equation:

$$q = q_0 + hp_{net} \tag{8}$$

The failure envelope intercept q_0 is related to the total cohesion through:

$$q_0 = ch\cot\phi \quad \text{and} \quad h = \frac{6 sen\phi}{3 - sen\phi} \tag{9}$$

where the total cohesion c is given by:

$$c = c' + c_{app} = c' + s\tan\phi^b \tag{10}$$

Therefore, the cohesion intercept is given by two aliquots, *i.e.*, the effective cohesion (c') and the apparent cohesion ($c_{app} = s\tan\phi^b$), where the friction angle ϕ^b is representative of the increase in strength due to suction.

It follows that the failure envelope in terms of stress invariants q, P_{net} can be rewritten as:

$$q = q_1 + sh\tan\psi^b + hp_{net} \tag{11}$$

where $\tan\psi^b = \tan\phi^b/\tan\phi$ and $q_1 = c'h\cot g\phi$.

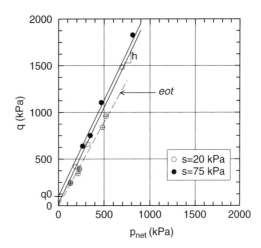

Figure 13. Sample at the end of test.

Figure 14. Failure envelopes.

Table 3. Strength parameters.

s (kPa)	h	ϕ_p (°)	ϕ_{eot} (°)	ϕ^b (°)	ψ^b (°)	q_0 (kPa)	q_1 (kPa)	c' (kPa)	c_{app} (kPa)	c (kPa)
20	2.06	50	44	37	33	23	0	0	15	15
75	2.06	50	44	37	33	100	0	0	57	57

Figure 14 shows the stress states at peak deviator stress and at the end of test (*eot*). Open and full circles identify tests at values of suction of 20 and 75 kPa respectively, while cross symbols refer to the stress states at the end of test. The analysis of data leads to the shear strength parameters reported in Table 3. The peak friction angle ϕ_p is found to be independent on the applied suction, while the total cohesion is mostly due to suction ($c' = 0$) and increases with it.

The stress states at the end of test plot close to a single straight line through the origin *i.e.*, the effect of suction on ultimate strength appears to be negligible. A least squares best fit of data obtained from all tests results in a ultimate failure envelope of equation $q = 1.79 \, p_{net}$, corresponding to a friction angle $\phi_{eot} = 44°$.

6 CONCLUSIONS

The hydraulic and mechanical properties of partially saturated *Pozzolana Nera* (Roma, Italy) have been presented. The experimental soil water retention curve has been obtained from volume pressure plate extractor tests.

Different approaches have been adopted to estimate the conductivity function of the material (analytical methods, *e.g.* Gardner, 1956; Rijtema, 1959; semi-empirical, *e.g.* Green & Corey, 1971; statistical methods, *e.g.* Van Genuchten, 1980).

The results of oedometer/wetting tests showed that the material, when saturated, exhibits a "collapsible" behaviour, whose amount depends on initial voids ratio, degree of saturation and stress level.

Data obtained from triaxial tests with shearing at constant suction allowed to define the stress–strain behaviour of the material. Both the peak shear strength and the initial tangent stiffness increase with increasing suction. A strong dilatant behaviour was observed.

Particular attention was focused on defining the failure envelope – according to an extended Mohr–Coulomb failure criterion – and assessing the influence of suction on such conditions. The peak friction angle is found to be independent on suction, while the total cohesion increases with it; the effect of suction on ultimate strength appears to be negligible.

ACKNOWLEDGEMENTS

The oedometer test denominated EDO5 was carried out at University of Roma *La Sapienza*. Tests EDO10 and TX15 were performed by Miss V. Papi at University of Perugia.

The Authors gratefully acknowledge the help and the suggestions received by Dr. M. Nicotera and Dr. L. Olivares during the set up of the triaxial system.

REFERENCES

Aversa S., Nicotera M. 2002. A triaxial and oedometer apparatus for testing unsaturated soils. *Géotechnical Testing Journal*, GTJODJ, Vol. 25, n. 1: 3–15.
Bishop A. W. (1959). The principle of effective stress. Lecture delivered in Oslo, Norway, in 1955; published in Tecnick Ukeblad, Vol.106, n. 39: 859–863.
Burdine N. T. 1952. Relative permeability calculations from pore size distribution data. *Trans. AIME*.
Cattoni E. 2003. *Comportamento meccanico e proprietà idrauliche della Pozzolana Nera dell'area romana in condizioni di parziale saturazione*. Ph.D. Università di Perugia.
Cecconi M. 1999. *Caratteristiche strutturali e proprietà meccaniche di una piroclastite: la Pozzolana Nera dell'area romana*. Ph.D. Thesis, Università di Roma *Tor Vergata*.
Cecconi M. Viggiani GMB. (2000). Stability of sub-vertical cuts in pyroclastic deposits, *Geoeng 2000 International Conference on Geotechnical & Geological Engineering*, Melbourne, Australia, 19–24 Nov.
Cecconi M., Viggiani GMB. 2001. Structural features and mechanical behaviour of a pyroclastic weak rock, *Int. J. Numer. Anal. Meth. Geomech* 2001; 25: 1525–1557.
Cecconi M., De Simone A., Tamagnini C., Viggiani GMB. 2002. A constitutive modelling for granular materials with grain crushing and its application to a pyroclastic soil, *Int. J. Numer. Anal. Meth. Geomech* 2002; 26: 1531–1560.
Cecconi M. De Simone A., Tamagnini C., Viggiani GMB. 2003. A coarse grained weak rock with crushable grains: the Pozzolana Nera from Roma, *International Workshop on "Constitutive Modelling and Analysis of Boundary Value Problems in Geotechnical Engineering"* – 3X4", Napoli 22–24 April 2003, 157–216.
Cuccovillo T., Coop MR. 1998. The measurement of local strains in triaxial tests using LVDT's. *Geotéchnique* 47 (1): 167–172.
Fredlund D. G., Rahardjo H. 1993. *Soil mechanics for unsaturated soils*. Wiley & Sons, Toronto, 517.
Fredlund D. G., Xing A. 1994. Equations for the soil-water characteristic curve. *Can. Geotech. J.* Vol. 31: 521–532.
Gardner W. R. 1956. Calculation of capillary conductivity from pressure plate outflow data. *Proc. Soil Sci. Amer. Soc.*, Vol. 20, pp. 317–320.
Green R. E., Corey J. C., 1971. Calculation of hydraulic conductivity: a further evaluation of some predictive methods. *Soil Sci. Amer. Proc.,* Vol. 35: 3–8.
Kunze R. J., Uehara G., Graham K. 1968. Factors important in the calculation of hydraulic conductivity. *Proc. Soil Sci. Soc. Amer.*, Vol. 32, pp. 760–765.
Mualem Y., 1976. A new model for predicting the hydraulic conductivity of unsaturated porous media. *Water Resour. Res.* Vol. 12: 513–522.
Nicotera, M. 1998. *Effetti del grado di saturazione sul comportamento meccanico di una pozzolana del napoletano*. Ph.D. Thesis, Università di Napoli Federico II.
Nicotera M., Aversa S., 1999. Un laboratorio per la caratterizzazione fisico-meccanica di terreni piroclastici non saturi. Atti del XX Convegno Nazionale di Geotecnica, Parma, 201–212.
Nielsen R., Jackson D., Cary J. W., Evans D. D., 1972. Soil water. *Amer. Soc. Agronomy and Soil Sci. Amer.*, Madison, WI.
Oldecop L. A., Alonso E. E., 2001. A model for rockfill compressibility. *Géotechnique*, Vol. 51, n. 2: 127–139.
Richards L. A., 1931. Capillary conduction of liquids through porous medium. *J. Physics*, Vol. 1: 318–333.
Rijtema P. E., 1959. Calculation of capillary conductivity from pressure plate outflow data with non-negligible membrane impedance, Netherlands, *J. Agri. Sci.,* Vol. 14: 209–215.
Van Genuchten M. Th., 1980. A closed-form equation for predicting the hydraulic conductivity of unsaturated soils. *Soil Sci. Soc. Am. J.* 44: 892–898.

On the suction and the time dependent behaviour of reservoir chalks of North Sea oilfields

G. Priol, V. De Gennaro, P. Delage & Y.-J. Cui
Ecole Nationale des Ponts et Chaussées (CERMES, Inst. Navier, ENPC-LCPC), Paris, France

ABSTRACT: In the North Sea Ekofisk oilfield, oil is contained in a 150 m thick layer of porous chalk (n = 40–50%) located at a 3000 m depth. An enhanced oil recovery procedure carried out by injecting sea water in the reservoir chalk caused a significant subsidence resulting in an about 10 m settlement of the seafloor. This coupled multiphase problem has been considered within a framework taken from the mechanics of unsaturated soils, by replacing the non wetting fluid (air in unsaturated soils) by oil in the reservoir chalk. This approach lead to consider the oil-water suction $s = u_o - u_w$ (where u_o and u_w are the oil and water pressures, respectively) as a relevant independent stress variable to describe the coupled behaviour of reservoir chalks. Another important characteristic of chalk behaviour is the time dependency of its constitutive law. In an attempt to better understand the subsidence observed under waterflooding, the objective of this paper is to investigate the combined effect of suction and time on the response of the multiphase chalk under isotropic compression. The result obtained confirm that the LC concept of the Barcelona model is also relevant for reservoir chalk. The combined effect of suction and loading rate on the shape of the LC curve is evidenced, providing a continuous link between existing data on the behaviour of water saturated and oil saturated chalk samples.

1 INTRODUCTION

A 10 m seafloor subsidence due to sea water injection within reservoir chalks has been observed in the Ekofisk North Sea oilfields during the last 30 years. The present subsidence rate of 0.4 m per year remains quite significant. Subsidence has generated significant extra costs and made necessary the elevation of the Ekofisk offshore platforms.

Subsidence is due to the compaction of chalk layer during seawater injection. A coupled multiphase analysis based on the framework of unsaturated soils mechanics was proposed by Delage et al. (1996). In this approach, oil is considered as the non-wetting fluid (as air in unsaturated soils) and water as the wetting fluid (as in unsaturated soils). Figure 1a presents a schematic illustration of this approach, together with two scanning electron microscope (S.E.M.) photos (1b) of Lixhe chalk, used in this study. Subsidence due to water injection was interpreted as a phenomenon of collapse due to wetting under load, typically observed in loose unsaturated low plasticity soils.

Like in unsaturated soils, the oil-water suction $s = u_o - u_w$ (where u_o and u_w are the oil and water pressures, respectively) was considered as a relevant independent stress variable to investigate the mechanical behaviour. Besides capillary actions, the suction also includes the possible physico-chemical interaction existing between chalk and water. This approach that is complementary to standard investigations already carried out with only one fluid (either water or oil, Ruddy et al. 1989, Schroeder et al., 1998, Risnes and Flaageng 1999, among others) appears to be necessary to better account for the problem.

In this paper, the experimental systems used for the study of the multiphase chalk are introduced, and preliminary results highlighting the combined effects of suction and time on the isotropic compressive behaviour of chalks are presented. This study is part of two EU founded projects: Pasachalk1 & Pasachalk2 (1997–2003).

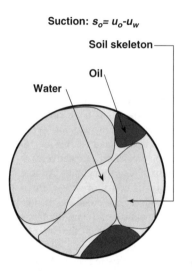

Figure 1a. Simplified scheme of the multiphase chalk.

Figure 1b. SEM photograph of Lixhe chalk (Pasachalh1, 2001).

2 MATERIALS

2.1 *Chalk*

Chalks are friable biomicrites mainly built up of skeletal debris of unicellular plankton algae called coccolithophorids. Coccolithophorids are often disaggregated into non-articulated grains that are actually crystals of calcite, with an average grain size range from 0.5 to 10 μm (see Figure 1b).

Due to the significant cost of field plugs, investigations are commonly performed on outcrop chalks that have never been in contact with oil. This shortcoming may have some important implications in terms of behaviour. The tests presented in this paper have been carried out on outcrop chalk coming from a quarry near Lixhe (Belgium). Geologically, Lixhe chalk originates from the upper Cretaceous period (between the Maastrichian and the Campanian periods), similar to that of the Ekofisk field, called "HOD Formation" in the Norwegian oil-bearing denomination (Andersen 1995). It originates from pelagic deposition in very quiet, slow and undisturbed conditions. The depth of deposition ranges between 100 and 600 m (300 in the Ekofisk case) and sometimes exceeds 1 000 m. The depositional mechanism is the main factor managing the final composition and behaviour of the rock (Clayton 1983).

Lixhe chalk is a pure white chalk with less than 1% of silica and with an average porosity of about 43%. Previous permeability measurements (Pasachalk1, 2001) provided a value of the intrinsic permeability of about $K = 1 \times 10^{-14}$ m^2 ($k_{water} \cong 1 \times 10^{-8}$ ms^{-1}, $k_{oil} \cong 7 \times 10^{-9}$ ms^{-1}).

2.2 Preparation of the chalk samples

Due to ductile behaviour, special care was devoted to chalk sample preparation. Standard triaxial samples (76 mm height and 38 mm diameter) were shaped on a lathe at the required size, equalising the lateral and the horizontal surfaces. Specimens were afterwards dried at 105°C for 24 hours in order to eliminate most of the water. Lord et al. (1998) showed that only free water is eliminated in such conditions, while some amount of structural water still remains within the chalk.

Samples were afterwards saturated under vacuum by the selected fluid (water or oil) for 24 hours. The sample masses were weighted before and after drying and after saturation in order to calculate the average porosity and the saturation degree.

2.3 Oil and wettability

The fluid called oil in this study is a non toxic, non polar organic liquid called Soltrol 170 (Phillips Petroleum Company). Soltrol 170 does not contain any polycyclic aromatic hydrocarbons, it has a very low water solubility ($\ll 1$ mg/l at 20°C), a low air volatility ($\ll 4 \times 10^{-2}$ mm^3/h at 20°C) and is not water miscible. The Soltrol 170 dynamic viscosity is $\eta_{oil} = 2.028$ cP and the density is $\rho_{oil} = 0.78$ Mg m^{-3}. Of course, chalk wettability depends on the fluids used, and should obviously be different with Soltrol as compared with crude oil. These changes in wettability obviously affect the oil-water retention properties of chalk.

3 EXPERIMENTAL DEVICES

Two techniques have been used for the control of the suction. The osmotic technique has been adapted for the control of oil-water suction and has been used in the preparation of samples to be tested in the triaxial apparatus. The overpressure technique (also called axis-translation technique) has been used in the high pressure triaxial apparatus and in special cells for the determination of the oil-water retention properties.

3.1 Osmotic technique

The osmotic technique for the control of suction has initially been developed by biologists and the technique has been later on adopted by soil scientists (Zur, 1966). The first adaptation to geotechnical testing was by Kassif and Ben Shalom (1971) in an oedometer for studying swelling soils. Subsequent work has been done on a standard triaxial apparatus by Delage et al. (1987). Some improvements concerning the circulation of the solution and the control of water exchanges have been proposed by Delage et al. (1992) and Dineen and Burland (1995).

In the osmotic technique, samples are put in contact with regenerated cellulose semipermeable membrane behind which is circulated an aqueous solution of big sized molecules of polyethylene glycol (PEG). Since PEG molecules are too large to cross the semi-permeable membrane, an osmotic pressure is generated through the membrane by the PEG solution, resulting in a suction that is a function of the PEG concentration (c_{peg}). Equilibrium between the sample and the solution is reached when the suction inside the sample is equal to that imposed by the PEG concentration. A calibration curve relating the PEG concentration to suction has been proposed by Williams and Shaykewich (1969) and corrected for membrane effects by Dineen and Burland (1995). In this work, the value of the PEG concentration c_{peg} has been obtained through the measurement of the refractive Brix index (Br) of the PEG solution, according to the following relationship (Delage et al. 1998):

$$s = \left(\frac{Br\sqrt{11}}{90 - Br} \right)^2 \qquad (1)$$

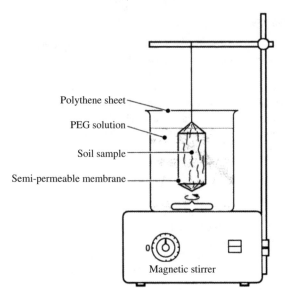

Figure 2. Schematic layout of the device for suction control using the osmotic technique.

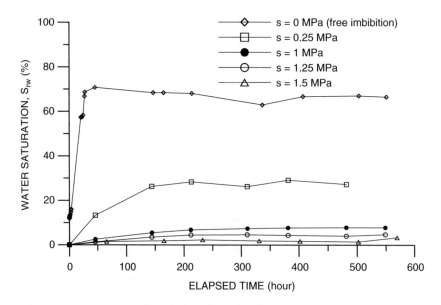

Figure 3. Imbibition curves of Lixhe chalk at various suction levels.

The osmotic suction has been used for the determination of the oil–water retention (or capillary pressure) curve of the chalk and to apply to triaxial samples the desired suction prior to triaxial testing. The method (Cui & Delage 1995, De Gennaro et al. 2003) consists in inserting the chalk sample in tube-shaped semi-permeable membranes and in plunging it into a container full of PEG solution placed on a magnetic stirrer (Figure 2).

For the determination of the water infiltration curve, the sample was saturated with oil before being inserted. Due to its wetting properties, water could infiltrate chalk under a controlled suction and expel a given volume of oil. Figure 3 shows typical trends giving the rate of infiltration observed when submitting the samples to various suctions ranging from 1.5 MPa down to 0 (free

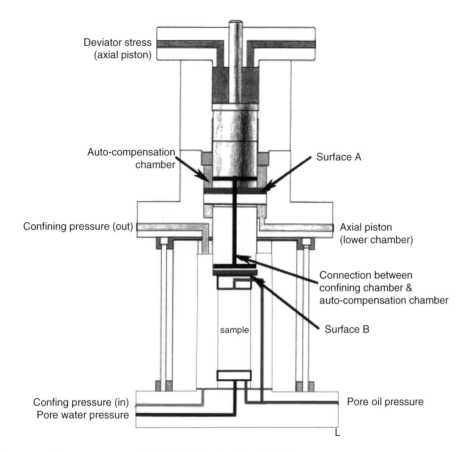

Figure 4. The auto compensated triaxial cell, designed by Geodesign.

water infiltration). It may be noted that less than 50 hours are necessary to reach equilibrium with free water, whereas up to 2 weeks are necessary with suctions higher than 1 MPa, with small amounts of water infiltrated.

Several precautions have to be taken when using regenerated cellulose semi-permeable membranes, regarding possible bacteria attacks and low resistance to compression and tearing. Several drops of penicillin were poured into the solution. It is also necessary to systematically change the membrane after a time period of 3–4 days.

3.2 *Triaxial testing device and procedure*

Suction controlled triaxial tests were carried out in a high pressure auto-compensated Geodesign triaxial cell (60 MPa maximum confining pressure, 100 MPa maximal deviatoric stress, Figure 5) with standard samples (38 mm in diameter and 76 mm in height). In this system, the axial load is applied by imposing a fluid pressure in the so-called auto-compensation chamber shown in Figure 4. Since surfaces A and B are equal, and thanks to a tubing connecting the confining chamber to the auto-compensation chamber, an increase in pressure in the confining chamber does not result in any axial upward effort on the piston. The confining and deviatoric stresses are applied by means of high pressure GDS pressure–volume controllers (64 MPa).

Oil and water pressures are separately controlled by using two standard GDS pressure–volume controllers (3 MPa). As in any standard axis translation device, the pressure of the wetting fluid (water) is controlled through a small cylinder shaped ceramic high air entry value (HAEV) porous

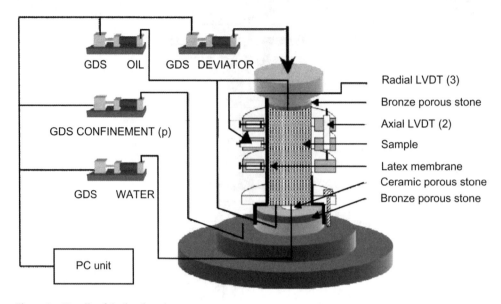

Figure 5. Details of the local strain measurements.

stone placed in the centre of the lower base, at the bottom of the sample (see also Figure 5). This porous stone is here impervious to oil. The oil pressure is controlled through a standard bronze porous stone placed on the top and through a hollow cylinder standard bronze porous stone that contains the HAEV porous stone in its centre, at the bottom of the sample.

Local radial and axial strain measurements were obtained by means of a special frame (Figure 5) mounted around the chalk sample and equipped with 5 LVDTs (Linear Variable Differential Transformers). It consists of three independent rigid rings equally distant one from another, fixed through the central ring around the middle height of the sample by means of three high stiffness springs oriented at 120° in the radial direction.

The average axial strain is measured by means of two axial LVDT (25 mm range) fixed to the upper ring that measure the relative displacement between the upper and the lower ring. The average radial strain is measured via the 3 horizontal LVDTs (5 mm range) oriented at 120° in the radial direction and mounted on the middle ring.

In the following, all volume changes observed during the isotropic tests have been calculated from the local strain measurements.

As commented before, samples were pre-equilibrated outside the cell using the osmotic technique and inserted in the cell at the desired suction. Subsequent suction control in the cell with the overpressure technique showed excellent agreement between the two techniques, with small further variations in the degree of saturation of water ($<1\%$). Specimens were mounted with filter papers placed at both ends and over the cylindrical surface in order to have a maximum drainage surface and a more homogeneous suction distribution.

4 OIL–WATER RETENTION PROPERTIES

Figure 6 presents the results obtained in terms of oil–water retention properties of Lixhe chalk using various techniques. The water drainage curve has been obtained by using two different techniques. The black dots correspond to experimental points obtained using the overpressure techniques in retention cells allowing for the independent control of oil and water, following a system similar to the one used in the triaxial cell described above. This points refer to the same sample, equilibrated at various suctions up to $s_o = 0.35$ MPa.

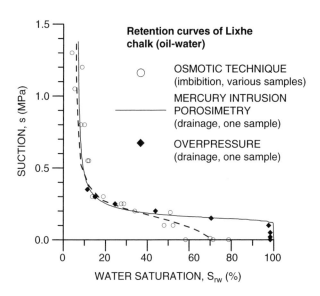

Figure 6. Oil–water retention curves of the Lixhe chalk.

The continuous line plot is the mercury intrusion porosimetry curve, obtained after a correction made by changing the interfacial tension values from the mercury/chalk values ($\theta = 141$–$146°$ and $\sigma_{hg-chk} = 480.10^{-3}$ N/m) to the water/chalk values ($\theta = 33$–$73°$ and $\sigma = 44.10^{-3}$ N/m). A good agreement is observed between these two approaches, showing that in the range of suction used using the over-pressure technique (up to 0.35 MPa), most of the retention phenomena are governed by capillarity. The drainage curve also indicates that a residual water degree of saturation S_{rw} of 5% is observed under 1.5 MPa suction. Note that the mercury intrusion curve (not presented here) exhibited a unique well graded pore population with an average entrance radius equal to 0.38 μm. Obviously, the significant change in water content observed between 0.11 MPa (oil entry value) and 0.3 MPa corresponds to the emptying of this pore population that corresponds to the inter-coccoliths pores observed using SEM in Figure 1.

The water infiltration curve obtained with the osmotic system (white circles and dot line interpolation, Figure 6) has been started with 5 samples full of oil at various suctions. Values of suction were defined using relation (1) and corrected based on Dineen and Burland's calibration (1995). A value of S_{rw} comprised between 5 and 10% is observed near 1.5 MPa. When suction is decreased (1.5, 1.25, 1, 0.75, 0.5, 0.25, 0.1 and 0 MPa), water is permitted to progressively infiltrate the samples by expelling the corresponding volume of oil.

At a zero suction, the water degree of saturation is comprised between 60 and 80%, which confirms the significant water wettability of Lixhe chalk. This is also the case of Ekofisk chalk, whereas other North sea reservoir chalks, such as Valhall chalk for instance, are not wettable to water but to oil. In this case, the intersection of the infiltration curve with the zero suction line would occur at a value smaller than 50%.

5 ISOTROPIC COMPRESSION TESTS

As in soils, the rate of application of strain or stress during compression tests run on chalk samples should be low enough to allow for pore pressure dissipation (Gibson & Henkel 1954). In unsaturated soils, a constant suction condition also requires low strain rate, as shown by Ho & Fredlund 1982 and Delage et al. 1987. Observations of literature data (Delage 2004) shows that axial rates of 1 μm/mn are generally adopted in triaxial testing of unsaturated soils.

Table 1. Characteristics of isotropic triaxial test samples.

Name	Diameter (mm)	Height (mm)	Dry mass (g)	Void ratio	Porosity (%)	Saturated mass (g)	Saturation degree (%)	Suction (kPa)	Rate
T1	36.01	75.40	122.71	0.683	40.6	153.86	100	0	Fast
T5	35.98	75.61	119.00	0.725	42.0	147.76	45	200	Fast
T7	37.88	75.97	134.49	0.700	41.0	161.99	2	1000	Fast
T2	36.13	75.71	121.33	0.721	42.0	146.09	0	1500	Fast
T4	37.90	70.85	123.83	0.723	42.0	152.21	30	200	Mod
T8	37.80	72.50	130.58	0.664	40.0	150.61	20	500	Mod
T9	35.74	74.34	115.71	0.734	42.3	146.59	100	0	Slow
T3	37.96	76.36	142.06	0.625	38.0	168.51	7	200	Slow
T6	37.83	75.86	133.87	0.701	41.0	162.10	15	1000	Slow

* Mod stays for moderate rate at 3.2×10^{-4} MPa/s.

Less data are available in chalks in this regard. Havmøller and Foged (1998) considered that axial strain rates of about 0.1%/h (i.e. 2.7×10^{-7} s^{-1}) were slow enough to avoid any excess oil-pressure generation on almost fully oil-saturated chalks ($S_{rw} \cong 5\%$). Deviator loading rates of 1.6×10^{-4} MPa s^{-1} have been applied during triaxial tests at constant confining pressure by Homand and Shao (2000) in order to prevent excess pore pressure. However in this case the induced volumetric strain rate was unknown.

In this work, all the tests presented have been run following a load controlled procedure. Based on previous findings, two isotropic loading rates have been adopted: 3.3×10^{-3} MPa/s (fast rate) and 5.5×10^{-5} MPa s^{-1} (slow rate). Table 1 shows the characteristics of the samples used for the 9 isotropic compression tests performed. Note that two of them have been performed under a moderate stress rate of 3.2×10^{-4} MPa/s. The table also gives the suction and the rate applied during the tests, in an attempt to investigate the combined effect of suction and time on the behaviour of chalk. As commented before, all samples were previously submitted to suction equilibrium at the desired value using the osmotic system of Figure 2.

5.1 Preliminary discussions

Since no measurements of water and oil pore pressures were performed during the tests, some conjectures arise on whether all these tests can be considered fully drained or not depending on the loading rate. The magnitude of undissipated pore-fluid pressures (i.e. oil and water pressures) is controlled by the permeability and the compressibility of the chalk and depends on the loading rate. Following the general approach typical of unsaturated soils, two independent stress variables describe the mechanical behaviour of water and oil saturated chalk, namely: the mean net stress $p_{net} = p - u_o$ and the oil-water suction $s = u_o - u_w$. Consequently, if excess pore pressures are supposed to develop, being $u_o > 0$, these results in decreasing values of mean net stress p_{net}, and higher values of mean total stress at yielding. Besides, excess pore-fluid pressures induced by fast loading will affect the value of the imposed suction, depending on the degree of saturation of chalk in both fluids, leading to a variation of suction before the re-establishment of the initial constant value. Chalk compressibility, fluids viscosity, surface wettability, continuity of the fluid phases and the existence of bulk water (low suction) or menisci-water (high suction) could be possible factors influencing the excess pore-fluid pressures generation/dissipation.

De Gennaro et al. (2003) performed isotropic compression tests on chalk samples at various suctions under load control. During tests, alternating loading at slow/fast rate and creep stages were performed. It was observed that the induced volumetric strain rate under isotropic compression at slow rate (i.e. 5.5×10^{-5} MPa s^{-1}), of about 1.5×10^{-7} s^{-1}, was almost the same during creep (i.e. while sample was strained under constant load). These findings seem to support the idea that rate of 5.5×10^{-5} MPa s^{-1} is slow enough to avoid excess pore fluids pressures generation. However,

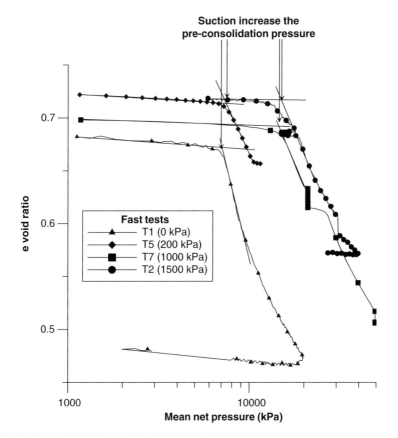

Figure 7. Suction influence on the isotropic compression.

when passing from a loading phase at fast rate (i.e. 3.3×10^{-3} MPa s^{-1}) to a creep phase and back, change in volumetric strain rate was noticeable, the maximum volumetric strain rate being of about 1.5×10^{-5} MPa s^{-1}. Tentative analyses based on consolidation theory (Bjerrum 1967, Janbu 1985, Leoureil et al. 1985), are hardly suitable in assessing the behaviour of such a soft rock, principally because no clear end of primary consolidation is detectable from consolidation curves. The analysis of the stress–strain-strain rate relationship for chalk seems to offer a viable way to have an insight into this aspect. This investigation is under consideration and is not discussed in this paper.

Based on results from oedometric tests it is believe that no significant pore pressure is generated when considering loading at the fast rate here assumed. This seems to be associated to the quite high permeability of chalk ($k_{water} \cong 1 \times 10^{-8}$ ms^{-1}, $k_{oil} \cong 7 \times 10^{-9}$ ms^{-1}) and the low compressibility of the soil skeleton, in agreement with an estimation of the Biot's coefficient b (where $\sigma = \sigma' - bu$) of 0.85 (Engstrom 1992).

5.2 *Experimental results*

5.2.1 *Suction effects*

Four isotropic compression test results are presented in Figure 7. Results confirm what already observed during oedometric test (De Gennaro et al. 2004). Increasing suction induces a progressive passage from a "water-like" behaviour to a "oil-like" behaviour. Generally this corresponds to a stiffer response in the elastic regime, an increase of the mean total stress at yield and a more pronounced transition from elastic to elastoplastic regime when yield occurs.

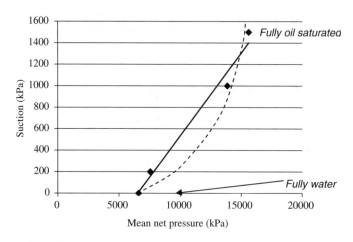

Figure 8. Suction effects (fast loading rate).

The fully oil saturated sample is the strongest and less compressible. Since suction of 1 MPa lies on the asymptotic limit of the retention curve shown in Figure 6, considering this asymptotic limit, we assume that oil saturated samples have a "fictive" suction of 1500 kPa. This suction is fictive because effective only if water is present.

Because the natural variability of the outcrop block porosity, samples have not the same void ratio. This is clearly shown in Figure 7. Chalk mechanical behaviour has a well proved sensitivity to porosity, and the elastic limit varies almost linearly with it. To prevent observations from porosity variability, based on experimental data, a coefficient of about 30 MPa per point of void ratio was assumed. The porosity influence is assumed to be unmodified by the loading rate and suction. This assumption could be questionable, as higher yield stresses are obtained for fast rate and oil saturated samples, It needs to be corroborated by further experimental data.

To highlight suction effects, according to the LC (Loading Collapse) locus presented by Alonso et al. (1990) for unsaturated soils, we have plotted in Figure 8 the mean net pressure at yield versus suction for fast loading rate tests. We can observe that the same chalk saturated by oil exhibits an elastic limit twice the water saturated chalk. The LC curve of Barcelona Basic Model seems then suitable in reproducing suction hardening of multiphase chalk.

5.2.2 Loading rate effects

Increasing the loading rate has obvious effects on the mechanical behaviour of chalk despite the porosity variability. Figure 9 highlights clearly this point. At the same suction level (1000 kPa), T7 was loaded faster than T6 test. We can observe that both pre-consolidation pressure and bulk modulus increase. The latter pass from a value of about 600 MPa to more than 2500 MPa, showing clearly that the elastic stiffness increases with loading rate. A similar increase is observed for the mean net pressure, that moves from 9 MPa to 14 MPa.

Results of the nine isotropic compression tests (Table 1) are summarised in Figure 10. Suction and loading rate have similar effects; both increase the elastic limit of chalk. However the intensity of this effect has to be related to the different shape of the LC curve in Figure 10. It looks likely that loading rate effects are more pronounced on fully oil saturated samples than on fully water saturated samples. Indeed, moving from a slow loading rate to a fast loading rate, mean net stress is multiplied by a factor of 1.25 in a water saturated sample and by a factor of 2.5 in a fully oil saturated sample. Admitting a linear LC locus, increasing loading rate increase the slope of the LC locus, in other words the yield locus for slow and fast rate are not parallel. This seems to offer some interesting information for further time-dependent modelling of multiphase chalk under suction controlled conditions.

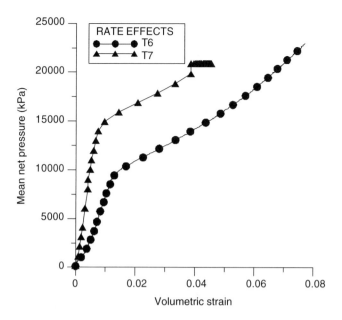

Figure 9. Effects of loading rate on 1000 kPa suction level tests.

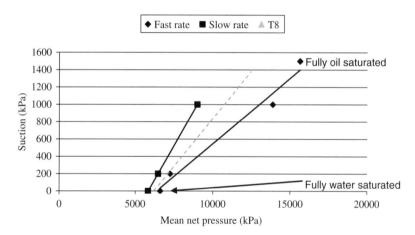

Figure 10. Yield stress function of loading rate and suction level.

6 CONCLUSION

Preliminary results obtained with both suction controlled triaxial apparatus permit us to draw general characteristics of the time dependent behaviour of multiphase oilfield chalk. First, they confirm the collapsible behaviour of oilfield chalk submitted to water injection, and the next stage would be the study of the collapse during an injection. Second, the loading rate effects have been studied and to a great extend identified. However, the larger breakthrough is the arisen possibility of a modification of the suction effects by loading rate. These last remarks warn us against comparing suction controlled tests at different loading rate despite a good drainage and good suction control.

REFERENCES

Alonso, E.E., Gens, I. & Josa, A. 1990. A constitutive model for partially saturated soils, *Géotechnique* 40(3): 405–430.

Andersen, M.A. 1995. Petroleum research in North Sea chalk. Joint chalk research phase IV: 47–153.

Bjerrum, L. 1967. Engineering geology of Norwegian normally-consolidated marine clays as related to settlement of buildings. *Géotechnique* 17: 81–118.

Clayton, C.R.I. 1983. The influence of diagenesis on some index properties of chalk in engineering. *Géotechnique* n°33–3: 225–241.

Cui, Y.-J. & Delage, P. 1995.

De Gennaro, V., Delage, P., Cui, Y.-J., & Schroeder, C. & Collin, F. 2003. Time-dependent behaviour of reservoir chalk: a multiphase approach. *Soil and foundations*.

Delage, P., Schroeder, C. & Cui, Y.-J. 1996. *Subsidence and capillary effects in chalk*, Proc. Eurock 96, Turin: 1291–1298.

Delage, P., Suraj De silva, G.P.R. & De Laure, E. 1987. *un nouvel appareil triaxial pour les sols non saturés*, 9th Eur. Conf. Of soil mechanics, Dublin: 26–28.

Delage, P. 2004. Experimental unsaturated soil mechanics: general report. *Proc. 3rd Int. Conf. on Unsaturated Soils UNSAT'2001*, Vol. 3, Recife, Balkema, in press.

Dineen, K. & Burland, J.B. 1995. A new approach to osmotically controlled oedometer testing. *Unsaturated Soils*. Rotterdam: Balkema.

Gibson, R.E. & Henkel, D.J. 1954. Influence of the duration on tests a constant rate of strain on measured drained strength, *Géotechnique* 4(1): 6–15.

Havmøller, O. & Foged, N. 1998. Confidential report DGI.

Ho, D.Y. & Fredlund, D.G. 1982. Strain rates for unsaturated soil shear strain testing. *Proc. 7th South East Asia Conf. Soil Mech. And Found. Eng.*, Hong-Kong: 787–803.

Homand, S. & Shao, J.F. 2000. Mechanical behaviour of a porous chalk and effect of saturating fluid. *Int. J. of Mechanics of Cohesive-Frictionnal Materials* (5): 583–606.

Janbu, N. 1985. Soil models in offchore engineering. *Geotechnique* 21(3) 25th Rankine lecture: 241–281.

Kassif, G. & Ban Shalom, A. 1971. Experimental relationship between swell pressure and suction, *Géotechnique* 21(3): 245–255.

Leroueil, S., Kabbaj, M., Tavenas, F. & Bouchard R. 1985. Stress strain strain-rate relation for compressibility of sensitive natural clays, *Géotechnique* 35(2): 159–180.

Lord, C.J., Johlman, C.L. & Rhett, D.W. 1998. Is capillary suction a viable cohesive mechanism? SPE/IRSM 47310 – Proc. Eurock 98: 479–485, Trondheim.

Risnes, R., Flaageng, O. & Madland, M.V. 2000. Some aspects of chalk-fluid interactions. *6th Chalk Symposium JCR V*. Brighton.

Ruddy, I., Andersen, M.A., Pattillo, P.D., Bishlawi, M. & Foged, N. 1989. Rock compressibility, compaction and subsidence in high porosity chalk reservoir: a case study of Valhall field. *J. of Petroleum Technology* 41(7): 741–746.

Schroeder, Ch, Bois, A.P., Maury, V. & Halle, G. 1998. Water/chalk (or collapsible soil) interaction: part II. Results of tests performed in laboratory on Lixhe chalk to calibrate water/chalk model. SPE/ISRM Eurock'98. Trondheim.

Williams, J. & Shaykewich, C.F. 1969. An evaluation of polyethylene glycol PEG 6000 and PEG 20000 in the osmotic control of soil water matric potential. *Can. Geotech. J.* 102(6): 394–398.

Zur, B. 1966. Osmotic control the matrix soil water potential, *Soil Science* 102: 394–398.

Experimental study on highly compressible neutralised and non neutralised residues exposed to drying

Lúcio Flávio de Souza Villar
Department of Transportation Engineering and Geotechnic, Engineering School, Universidade Federal de Minas Gerais – UFMG – Brazil

Tácio Mauro Pereira de Campos
Department of Civil Engineering, Pontifícia Universidade Católica do Rio de Janeiro – PUC-Rio – Brazil

ABSTRACT: This paper presents some results of tests performed in highly compressible materials submitted to drying. It is outlined the methodology used to obtain their SMCC, aiming the understanding of their behaviour when exposed to solar drying, an important information to achieve new disposal concepts and methods. The studied material was a silty clay waste from the alumina industry, known as "red mud". These materials usually have a caustic liquid (pH around 13) as pore fluid. Sometimes, to neutralize the residues before their disposal, sulphuric acid is added to them. Four different techniques were used to obtain the relationship between moisture and suction. They were: the filter paper method, small tensiometers installed directly in contact with the waste inside a drying box, a suction probe and osmotic desiccators. The methodology employed and the difficulties and advantages found using these techniques are presented. The pore fluid influence in some properties such as the SMCC, the tensile strength and drying curves is presented and briefly analysed.

1 INTRODUCTION

The development of geoenvironmental engineering is increasingly demanding knowledge on materials and situations that geotechnical engineers are not used to deal with. It is quite common that during some or the entire life of geoenvironmental projects, all the materials involved will be in an unsaturated state. Moreover, their implementation frequently occurs in arid or semiarid regions, where soils are unsaturated almost the whole time (Barbour 1998). The design of containment facilities, the reclamation of mines and industrial sites, cover design for landfills or mine wastes, contaminant transport through soil mass are examples of environmental situations involving these different materials. One consequence is the growing need for theories and new technologies to correctly predict their behaviour.

This research was motivated by the perception that practical problems comprising the disposal of industrial and/or mineral wastes are growing fast. It is quite common that these wastes are disposed as slurries and then let to settle inside reservoirs. Such methodology requires large areas to accommodate the usually high quantities of produced wastes, involving a number of environmental issues. In regions with an elevated level of solar incidence, an alternative is to let these materials to dry, reducing their volume and increasing the reservoir lifetime. But this technique, less aggressive to the environment, needs to be better investigated.

The research on the behaviour of wastes launched as slurry and then exposed to dry has many variables involved, what turns the problem complex and challenging. The reservoir life span is strongly influenced by the local hydrological cycle and the materials can experiment different processes and be in different stages at the same time. For instances, they can be unsaturated and completely fissured in some regions but saturated and still consolidating in another.

In an attempt to contribute to the understanding on the behaviour of residues disposed as slurries and the development of suction when they are exposed to solar drying, a large research programme was set up at PUC-Rio. The analysed materials were slurry wastes from the alumina industry, known as "red muds", and from bauxite washing after mining. The pore fluid of this last type of residue is water. However, in the case of the red muds, pore fluids varied from a very caustic liqueur, with pH above 13, to a neutralised one, resulting from mixture with sulphuric acid. The high pH is a consequence of the Bayer Process worldwide employed to obtain the alumina; the mixture with sulphuric acid being sometimes used to decrease this pH before waste disposal.

Five residues from different sites in Brazil were studied. The goal was to develop an adequate but simple and cheap methodology, able to describe with reliability their drying process, and designed to provide waste parameters to supply numerical models used to predict field situations comprising both their saturated disposal conditions and effects of subsequent drying.

To achieve this objective, the research programme encompassed different types of laboratory and field tests (Villar 1990, 2002). Laboratory tests included a special geotechnical characterization program and the determination of saturated parameters using constant rate of displacement and conventional oedometric consolidation tests. The Brazilian test was used to determine tensile strength on unsaturated samples. A laboratory simulation of solar drying process was developed and the fissuring pattern of the wastes investigated. Suctions were measured or achieved using different methods. Field tests in both saturated and unsaturated wastes were performed to obtain actual humidity, density and pore pressure profiles. Laboratory results were compared with those from field monitoring programs (e.g. de Campos et al 1998).

This paper outlines only the methodology used to obtain the soil moisture characteristic curve (SMCC) and some aspects related to the behaviour of two of the studied residues when submitted to drying. They comprise the same raw material, with different pore fluid. Some aspects of this methodology, detailed in Villar (2002), has been described in Villar & de Campos (2002). De Campos et al (1994) and Azevedo et al (1994) discussed aspects, not commented in the present work, related to the sedimentation/self weight consolidation behaviour of one of the residues (neutralised one).

It is interesting to note here that the expression "soil MOISTURE characteristic curve" (SMCC) instead of "soil-WATER characteristic curve" (SWCC) is used throughout the paper. This is because while this latter definition considers only water as pore fluid, the first one (SMCC) does not.

2 TESTED MATERIALS: CHARACTERIZATION RESULTS

Although five types of residues were studied in the research programme, only two, also described in Villar & de Campos (2003), are considered here. They consist of *red muds* resulting from the physical-chemical treatment of bauxite in alumina production plants from the Brazilian southeast region. This name, *red mud*, is due to their colour, a consequence of their high iron content, and the very low consistency they are disposed off. One of the muds, named *NM residue* (neutralised mud), was neutralised (pH around 8), while the other, named *NNM residue* (non neutralised mud) was saturated with a caustic liqueur (pH around 13).

As will be seen in the following sub-items, a number of aspects related to the characterization tests routines and data obtained merits further description and discussion. This, however, is outside the scope of the present paper, being presented elsewhere.

2.1 *Chemistry and mineralogy*

The red mud final composition depends upon the rock quality and the used processing technology (e.g. Bulkai 1983). Basically, all the residues conserve hematite and silicates from the original rock. Quartz and clay minerals are usually absent due the contact with the caustic soda used in the processing phase (e.g. Somogyi & Gray 1977). As a result of such processing, a sub product called *Bayer sodalite* may exist. This sub product affects the mud behaviour, especially its sedimentation

Table 1. Chemical results for the NM and NNM residues.

Material	SiO_2	Al_2O_3	Fe_2O_3	FeO	CaO	MgO	TiO_2	P_2O_5	Na_2O	K_2O	MnO	Cr_2O_3
NM	5.7	17.2	52.8	0.56	3.1	0.15	5.5	0.37	1.9	0.15	0.06	0.097
NNM	6.2	17.0	50.3	0.42	4.0	0.21	5.5	0.35	2.4	0.12	0.05	0.088

rate within a caustic soda media, turning it slower than in water (Li 2001). According to Li & Rutherford (1996), the composition and amount of amorphous material present in the residues may also influence its behaviour.

Table 1 summarises chemical results obtained for the two residues analysed here. As can be seen, from the chemical point of view, they are quite similar. Iron and alumina oxides predominate in the composition of both residues, followed by silica, titanium, calcium and sodium oxides. Results of X-Ray diffraction analysis showed, as expected, the absence of quartz and clay minerals in both residues. In the NNM residue it was, however, noticed the occurrence of alophane. This amorphous material probably does not occur in the NM residue due the neutralization process with sulphuric acid imposed to this mud.

In the drying process the muds were submitted to, it was observed both the occurrence of a material precipitated within the voids of the dried samples and the formation of a thin, white crust, on the face of the samples exposed to atmosphere. X-Ray analysis of these white materials, that appeared as a result of the evaporation of the liqueurs, indicated that they were composed by $Al(OH)_3$, which probably is an amorphous hydroxide of aluminium. In the case of the NM residue it was also identified the occurrence of *thernardite*, an element resulting from the combination of sodium and sulphuric acid, while in the NNM appeared a material classified as *trona*, a mixture of sodium, water and carbonic gas.

2.2 *Grain size distribution*

Usually, the grain size distribution of mining related residues depends upon the rock (bauxite) mineralogy and the technology used in the industrial processing. Both these features can vary from region to region and even from an industrial plant to another. The grain size distribution of materials originated from sedimentation processes, such as in the case here studied, is also strongly influenced by the type of fluid where deposition takes place. Ignatius & Pinto (1991) studied the effect of a caustic liquid, with a pH similar to that here analysed, in the dispersion of a clayey soil during sedimentation tests. They concluded that the particles could disperse or flocculate depending on the salt content in solution in the fluid. These authors observed that the grain size distribution curve obtained in tests using caustic liqueur tended to be quite different from that obtained when using water as sedimentation media, especially in the fine grain size range. In addition to such type of observations, the disposal technique of the muds can originate different grain sizes zones within the reservoirs, due to segregation effects. So, highly variable grain size distribution curves are likely to be associated to materials such as those here considered.

In the present study, before the grain size tests, the residues were air-dried and homogenised. Sedimentation tests were carried out with and without the use of sodium hexametaphosphate as dispersing agent and using both water and liqueur (caustic one in the case of the NNM residue and neutralised one for the NM residue) as sedimentation media. Materials passing in the sieves #40 (integral samples) and #200 (fines) were tested.

In all sedimentation tests performed using liqueur (caustic or not, with or without dispersant), it was observed that it crystallized, forming very fine plates that deposited at the bottom of the burette. This implied in a variation of the density of the fluid (order of 10%) during the tests, which was duly accounted for in the interpretation of the results.

Figure 1 shows grain size distribution curves of NM integral samples. Similar curves were obtained for the NNM residue. Table 2 shows lower and upper bounds of grain sizes considering all tests performed.

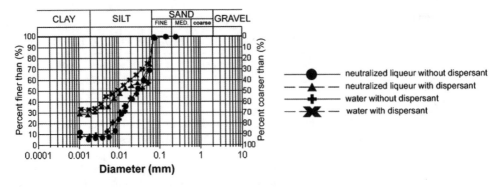

Figure 1. Grain size distribution curves for integral samples of the NM residue.

Table 2. Grain size distribution boundaries.

Residue	Neutralised mud			Non neutralised mud		
	% sand	% silt	% clay	% sand	% silt	% clay
Upper bound (water)	0	49	51	0	78	32
Lower bound (liqueur)	62	30	8	57	40	3

As expected, a large variation in grain size distribution was observed. The presence of sodium hexametaphosphate implied in the occurrence of larger amount of fines content both when water and liqueur were used as sedimentation media in the case of the NM residue. In the case of the NNM residue, it tended to act as a flocculating agent instead. Generally speaking, greater clay contents were obtained when sedimentation occurred under water. Thus, there was a trend of flocculation of the clay fraction in presence of both liqueurs, with such effect being larger in the NNM residue.

Further comments on the grain size distribution of the residues are outside the scope of this paper and will be presented elsewhere. However, in what concerns flocculation tendencies, it is interesting to mention here that observations under electronic microscope indicated that both residues tended to show flocculated clay particles after drying, with such flocculation being much higher in the NNM residue (Villar 2002). Also, it is interesting to note that, owing to the high dependence of the obtained grain size distribution on testing procedures, predictions of the SMCC based on grain size distribution (e.g. Fredlund & Xing 1994) should be seen with care in the case of these residues.

2.3 *Specific gravity*

The specific gravity of the residues is highly dependent on the chemical composition of the rock of origin. Table 3 show results obtained for the materials analysed here. Tests were performed using both liqueur and water inside the picnometers, and samples with different grain sizes.

As indicated in Table 3, both residues have high values of specific gravity, what can be explained by their high iron content (Table 1). The use of liqueur instead of water in the tests implied in lower values of specific gravity, the variations being more relevant in the case of the NM residue.

2.4 *Consistency limits*

As pointed out by Li & Rutherford (1996), although red muds may behave like high plasticity clays, they generally have low plasticity indexes. In agreement with that, as will be seen later, the residues here considered are highly compressible and, as shown in Table 4, have low plasticity.

The results shown in Table 4 refer to tests performed in samples that were initially air dried and than wetted, and samples where the tests started with the muds in their natural saturated state. In the

Table 3. Specific gravity of the residues.

Material		>#60	>#100	>#200	<#200	Integral
NM	Water	3.66	3.65	3.65	3.57	3.68
	Liqueur	3.49	3.49	3.51	3.54	3.39
NNM	Water	3.56	3.59	3.58	3.62	3.59
	Liqueur	3.53	3.60	3.54	3.54	3.55

Table 4. Liquid and plastic limits of the analysed residues.

		<#200			<#40			Integral - drying path		
Material		w_L	w_P	PI	w_L	w_P	PI	w_L	w_P	PI
NM	Water	38.5	27.1	11.4	41.4	32.7	8.7	–	–	–
	Liqueur	35.8	28.6	7.2	33.4	27	6.3	48.8	34	14.8
NNM	Water	38.5	27.1	11.4	41.4	32.7	8.7	–	–	–
	Liqueur	36.5	28.2	8.4	33.9	26.8	7	63.6	29.3	34.3

case of the initially air dried samples, tests were performed in the material passing in the sieve #200, besides the conventional ones that are performed in the material passing in the sieve #40. When performing tests following a wetting path (initially air-dried samples and adding water or liqueur to them), the two residues presented similar values of liquid and plastic limits. But, when performing tests following a drying path (using initially wet samples and than left them to dry), the residues showed an index of plasticity much higher than those obtained in the former way. This indicates the little capacity of re-hydration of these materials, especially in the case of the NNM residue.

The shrinkage limit of the NM and NNM residues was obtained following the ASTM D427 recommendations (air dried material passing in the #40 sieve), but using either water or liqueur (caustic or neutralized) as pore fluid. For the NM residue, the shrinkage limits thus obtained were typically higher than their corresponding plastic limits, with values ranging from 30 to 35. For the NNM residue, most of the results varied between 34 and 39, being always higher than the corresponding plastic limits. As the shrinkage limit is expected to be smaller than the plastic one for a given material, the results obtained following American standards (that are similar to Brazilian ones) do not appear to be correct. On the other hand, as will be seen in the next section, if the shrinkage limit is obtained from drying curves, as suggested by British standards (BS 1377), reasonable results are obtained.

3 SHRINKAGE CHARACTERISTIC CURVES

The soil volume variation due to shrinkage is usually represented by a void ratio-gravimetric moisture content relationship, called soil *shrinkage characteristic curve* – SCC (e.g. Bronswijk 1988). When submitted to a drying process, the same three different phases of shrinkage defined for soils can be identified in the SCC of a given residue (e.g. Oliveira Filho 1998). A linear phase, named *normal shrinkage phase*, occurs at higher moisture contents, when the material is saturated. Following that occurs a non linear zone. It is called *residual shrinkage*, and indicates a reduction of the rate of material volume variation and an increase in air volume inside it. The last phase, named *zero shrinkage*, occurs when the material does not contract anymore. Olsen & Haugen (1998) stated that, in structured soils, occur a fourth phase, in which the macropores are drained without any volume reduction. They called this phase as *structural contraction* and affirmed that remoulded soils would not present it. Owing to the type of residue here considered, this fourth phase is unlikely to be of interest.

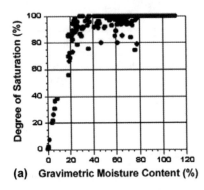 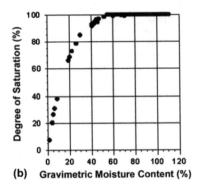

Figure 2. Drying curves for the NNM residue (a) before and (b) after data treatment.

3.1 Testing procedure

Shrinkage characteristic curves were obtained for both NM and NNM residues from analysis of systematic volume and moisture content measurements during drying tests performed as described by Villar et al (1997) and Villar & de Campos (1999). The main drying tests were executed in residues disposed as slurries into two cubical glass boxes with 0.5 m sides. The residues were left to consolidate by self-weight during at least a week after being well homogenised inside the boxes. Then, they were exposed to dry under special lamps that simulate solar radiation, as described by Swarbrick (1992). Such lamps were kept turned on for at least eight hours a day. The drying boxes were installed over mechanical balances. Consequently, it was possible to monitor the lost of weight during all the evaporation process. A stationary piston sampler with small diameter (1.9 cm) was used to collect samples from one of these boxes, weekly (the other one was left to dry without any mechanical disturbance). Immediately after collection, the samples were measured and weighted.

3.2 Data treatment

After computation of the index properties of the residues, curves relating them, called *drying curves*, were plotted in order to verify their inter-relationships and define the data that would better define the shrinkage behaviour of each material. Figure 2 shows an example of such drying curves, in which degree of saturation and gravimetric moisture contents are correlated. In Figure 2a all data obtained has been plotted, while Figure 2b shows the relationship considering only the results after data treatment. By examining drying curves such as that in Figure 2a, with different index properties being plotted in each axis, it was possible to eliminate data that did not fit general trends observed in each relationship. In other words, only those measurements that fit well all different relationships were considered, regardless if some of them provided, individually, good matches. In such way, potential errors associated to both volume and gravimetric water content measurements, resulting in inaccuracies in void ratio; volumetric water content and degree of saturation computations, specially when the samples were with higher moisture content, were minimised. Such data treatment methodology proved to be quite useful especially to determine relationships concerning the degree of saturation and volumetric water content. These indexes showed to be the most sensitive to eventual errors in the measurement of samples dimensions. For instances, a difference of circa 3% on the height of the collected samples could result in an error near, or even above, 10% in the computed degree of saturation, and of the order of 30% in the volumetric water content.

3.3 Obtained drying curves and discussion

Figure 3 shows the shrinkage characteristic curves obtained for the NM and NNM residues. Such curves were obtained considering the specific gravity of the grains of each residue as being equal,

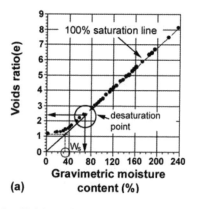 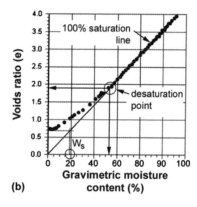

Figure 3. Shrinkage characteristic curves for the (a) NM and (b) NNM residues.

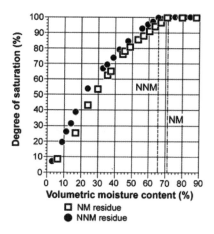 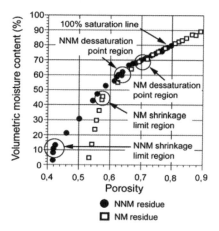

Figure 4. Degree of saturation–volumetric water content drying curves for the two residues.

Figure 5. Characteristic drying curves for the NM and NNM residues.

respectively, to the values obtained in the tests of integral samples using neutralized and caustic liqueurs ($G_s = 3.39$ and 3.55, see Table 3).

The 100% degree of saturation lines shown in Figure 3 correspond to the normal shrinkage phase of the residues. The end of this phase, or corresponding beginning of the residual shrinkage phase, defines the point were desaturation starts to occur in the residues. This point can be accurately defined in a drying curve as plotted in Figure 4. For the NM residue, the desaturation point occurs at a gravimetric moisture content of 68% (void ratio $= 2,4$ and volumetric water content $= 70,6\%$) while for the NNM residue, it occurs at a gravimetric water content of 54% (void ratio $= 1,9$ and volumetric water content $= 65,5\%$). As will be seen later, these moisture contents, obtained respectively from Figures 3 and 4, are those corresponding to the air entry value pressure, obtained from the SMCC related to each residue.

As mentioned earlier, before starting the drying tests the residues were left to consolidate under their own weights. Therefore, the large variations in void ratios shown in Figure 3 are only due to the drying process imposed to them, being indicative of the high volumetric compressibility of the residues when left to dry. In the saturated state, the NM residue showed a void ratio variation of 5.8, while for the NMM residue this variation was of 2,1. As the NM residue was disposed in the drying box with a solids content much smaller than in the case of the NNM residue, this difference in void ratio variation may be, at least partially, due to that. In other words, it can not be only due to the differences in the pore fluids and its effects in the microstructure of the residues. Both residues showed the same variation in void ratio under unsaturated condition ($\Delta e = 1,2$), with the NM residue reaching a minimum void ratio (zero shrinkage phase) of 1,2 and, the NNM, of 0,7. This difference in minimum void ratios reached under drying may, at least partially, be due the lower solids content of the NM residue.

From Figure 3, the shrinkage limits of the NM and NNM residues were found to be, respectively, equal to 32% and 19,5%. This limit is defined as equal to the gravimetric moisture content corresponding to the intersection of the lines defining the normal and zero shrinkage phases in the shrinkage characteristic curve. While the value obtained for the NM residue agrees well with that obtained following American standards (see section 2.3), that related to the NNM residue was much lower, being now compatible with the plastic limit of this material (Table 4).

Figure 5 shows another way to plot results of drying tests, which proved to be the most interesting in the present study. In the upper part of the volumetric water content x porosity drying curve, named here as *characteristic drying curve* (CDC), occurs a 45° inclination line. At the end of this line, it is possible to accurately define the point corresponding to the beginning of desaturation of the material. Another straight line occurs at the bottom of the CDC, the beginning of which coincided with the point corresponding to the shrinkage limit of each residue.

4 TENSILE STRENGTH

Little attention has been paid in the past on the tensile strength of soils, probably because it is widely propagated that soils do not have resistance to traction. While this may be true in relation to sedimentary saturated soils, it is not so in the case of unsaturated materials. In the research project into which this paper is included, the variation of the tensile strength with the degree of saturation was investigated. Partial results of such investigation are presented here.

4.1 *Testing procedure*

The tensile strength of the residues was obtained using the Brazilian test, largely employed to define tensile strength of concrete samples. Considering both soils and rocks, a number of authors have already described this type of test (e.g. Krishnayya & Eisenstein 1974; Maciel 1991; Favaretti 1995), whose theoretical aspects have recently been evaluated by Lavrov & Vervoot (2002).

The test consists in applying a compression stress in a point of the perimeter of a thin cylindrical (round *biscuit* shape) sample. By doing so, rupture is expected to occur by traction along the diameter of the transversal section of the sample.

After letting the residues to consolidate under their own weight, samples with diameter of 76 mm and 20 mm in thickness were moulded and than left to dry under air-conditioned conditions. After that, some samples were oven dried at a temperature of 105°C for one hour or so. Following this procedure, NM residue samples with gravimetric moisture contents ranging from 54% to 0.5% were obtained, while the moisture content of the NNM residue samples varied from 45% to 0.6%. Such moisture contents were obtained at the end of the tests, from material retrieved around the failure planes formed.

Following Maciel (1991), the samples were sheared in a triaxial compression machine at a constant rate of displacement of 2 mm/min. Two sets of samples were considered: those with no visible fissure (intact samples) and those with visible ones. Such fissures, observed in some cases, resulted from the drying process imposed to the sample prior to the test.

4.2 *Testing results and discussion*

For both materials, *intact* samples presented curves relating force and displacement with a clear peak, at which it was defined the tensile strength. The pre-fissured samples generally presented more than a single peak, the first one probably corresponding to the moment the preexisting fissure started to open. In these cases, the maximum monitored force was used to compute the tensile strength.

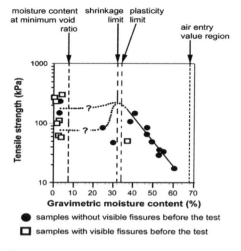
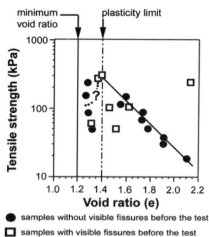

Figure 6. Tensile strength and gravimetric moisture content relationship for the NM soil.

Figure 7. Tensile strength and void ratio correlation for the NM soil.

Figures 6 and 7 show relationships between tensile strength and, respectively, gravimetric moisture content and void ratio, obtained for the NM residue. As can be seen in these Figures, the value of tensile strength increases while the gravimetric moisture content and void ratio values diminishes until reaching values corresponding to the plastic or shrinkage limits of the material. Up to moisture contents and void ratios corresponding to these limits, a straight line occurs in the plots, which have the tensile strength in a logarithm scale. This line goes up to a tensile strength of circa 250 kPa. From then on, either a constant tensile strength or a decrease of it can occur. In other words, no clear trend of behaviour was possible to be identified. Considering however that for moisture contents bellow the plastic limit the soil may become fissured and, in presence of fissures, the tensile strength would decrease, it is believed that the tensile strength of the residue decreases after reaching a peak value at a moisture content within the plastic and shrinkage limits. Findings by Maciel (1991), testing residual soils, and Favaretti (1995), testing compacted clays, support this view.

The occurrence of pre-existing fissures was not always easily noted at the surface of the samples. In this aspect, analysing relationships between tensile strength and different soil index properties showed to be helpful to identify fissured samples. For example, in Figure 7, it can be noted that when the sample was not intact before the test, its result does not belong to the existing correlation between the tensile strength and void ratio. In view of the fact that crack density may be a function of the sample size, it is supposed that the dimensions of the sample may influence the tensile strength test response, with such influence being larger for moisture contents at or bellow the plastic limit of the material.

The results obtained for the NNM residue are shown in Figure 8. It can be seen that they are very similar to that of the NM residue. The tensile strength showed the same tendency to continuously increase while the gravimetric moisture content decreases until near the shrinkage limit. After that, there is no clear tendency, but it was supposed that the tensile strength would decrease, due to crack formation, until a residual value. Comparing the NNM and NM results, it can be observed that both residues showed nearly the same maximum value of tensile strength. This indicates that differences in the pore fluid did not influence this residue property.

5 SOIL MOISTURE CHARACTERISTIC CURVES

For both residues, soil moisture characteristic curves were obtained after letting them to consolidate under their own weight until reaching consistencies or water contents permitting the retrieval of

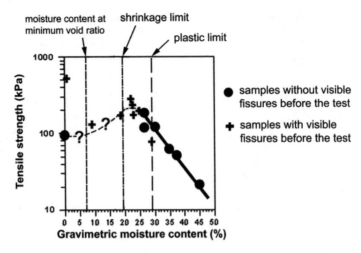

Figure 8. NNM residue tensile strength and gravimetric moisture content relationship.

samples with a minimum of mechanical disturbance. The samples were then left to dry in an air-conditioned environment. Thus, the SMCC here considered were obtained following a drying path.

5.1 Testing techniques

Four different techniques, already mentioned in Villar & de Campos (2002), and briefly described here, were used to obtain the soil moisture characteristic curves of the residues.

5.1.1 Filter paper

The filter paper technique is relatively well known (e.g. Houston et al 1994). It is quite simple and can provide measurement of both matric and total suction, depending if there is or not contact between the paper and the material being tested. The Whatman 42 filter paper was used, in conjunction with samples of 50 mm diameter. The calibration curve adopted was that suggested by ASTM D 5298-92. Special devices, described in Villar & de Campos (2002), were developed for the simultaneous measurement of both total and matric suction. With them it was possible to always guarantee a good contact between the filter paper and one of the sides of the sample, and a very small, always constant, gap between a second filter paper and the other side of the sample. Such devices were kept inside boxes placed in a room with constant temperature (20°C). Equalization periods varied from 7 to 120 days, with most of the tests lasting 15 days.

Finishing the equilibrium period, the chamber was disassembled near a 10^{-4} g precision balance. The filter papers were then placed inside capsules and their weights were monitored for at least five minutes. The weight data was then plotted against the square root of time to get the mass of the wet paper at the moment the container was opened (time = 0). This procedure was repeated to get the dry mass of the filter papers after oven drying, at a temperature around 105°C, during 24 hours.

5.1.2 Suction probe

Ridley & Burland (1993) presented a new device able to provide direct measurements of matric suctions up to 1500 kPa. This device, named *IC tensiometer*, was used in the research work only with the NM residue.

To perform the measurements, samples with 76 mm diameter and 20 mm height were prepared. The maximum gravimetric moisture content that was possible to do so was around 70%. After saturation of the probe, it was put with its porous stone upwards, on a levelled base. The sample was then placed lying on top of the stone. A saturated kaolin paste with gravimetric moisture content around 57% and thickness of circa 3 mm was used to improve the contact between the probe and

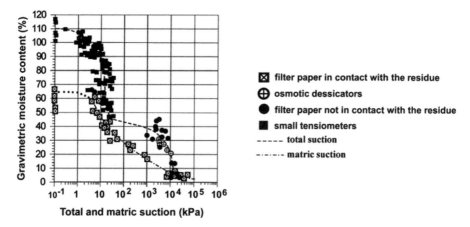

Figure 9. Suction and gravimetric moisture content relationship obtained plotting all data from the different utilized techniques – NNM residue.

residue. A plastic film was used to cover this ensemble to avoid moisture loss. Monitoring of probe readings was stopped when they reached an equilibrium value continuously maintained at least during 15 minutes. The moisture content of the samples was taken at the end of the readings. They ranged from 68% to 9%. In some tests, suction values up to 2300 kPa were measured. Typically, however, the probe lost its saturation at a pressure around 1200 kPa.

5.1.3 *Osmotic desiccators*
Osmotic desiccators were used to apply high levels of total suction. Inside them, the samples were left to equalize their vapour pressure with saline solutions of known molarity. They were placed as close as possible over the liquids, but with no contact. Following Juca (1990), solutions of NaCl in distilled water, able to apply suction levels varying from 1 to 10 Mpa, were employed. The samples were left inside the desiccators for approximately six months, being periodically weighted to verify if they had achieved constant mass.

5.1.4 *Mini-tensiometers*
Tensiometers of small diameter (5 mm), installed in the instrumented drying box referred to in section 3.1, provided suction measurements during the main drying tests performed. Such minitensiometers (10 unities, installed at different depths, in different points) were built with porous stones of air entry value of circa 100 kPa. Mercury columns were used to monitor suction variations. Thermocouples, installed approximately 2 cm apart from each tensiometer, provided measures of temperature variations and evaluation of their effects on the tensiometers readings (e.g. Warrick et al 1998). These tensiometers worked well up to suction values of 30 kPa. Above such level of suction, it was very difficult to keep them properly saturated.

5.2 *Results and discussion*

Figure 11 shows the relationship between suction and gravimetric moisture content obtained for the NNM residue. Equivalent results were obtained for the NM residue, as shown by Villar & de Campos (2002) and in Figure 10, that shows relationships between suction and, respectively, volumetric water content and degree of saturation for both residues.
 In Figures 9 and 10, the data registered by the mini-tensiometers are considered as being a measure of total suction. In a first analysis, it was thought that, as expected, such data were related to the measure of matric suction (Villar & de Campos 2001). However, further results provided by filter paper tests, as well as a careful re-evaluation of physical indexes fitting drying curves

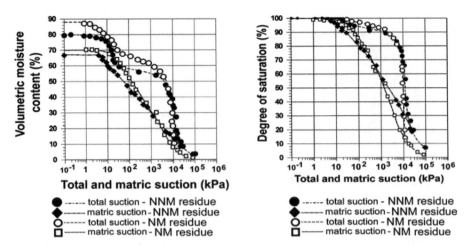

Figure 10. Suction × volumetric water content and degree of saturation relationships obtained plotting all data from the different utilized techniques.

obtained in the large drying box tests, led to the conclusion that the mini-tensiometers measured total suction instead of matric ones.

Regardless if the residues were neutralized or not, their pore fluid had high salt content. Thus, the occurrence of osmotic suction in them was expected. As the mini-tensiometers provided suction measurements essentially while the residues were still saturated, and the maximum capillary pressure associated to the drying boxes would be smaller than 5 kPa, it is apparent that such instruments provided a measure of total suction, function of the salt concentration existing in the residues while saturated. This indicates that the porous stone of these tensiometers may present some osmotic efficiency (e.g. Mitchell, 1991; Marinho, 1994).

For the entire range of the analysed moisture contents, the waste experienced a large volume contraction. This was true especially for values of volumetric water content varying from 85% to 50%. A possibility that must be verified is if this huge contraction could affect the behaviour of the small tensiometers and justify the appearance of the characteristic curves in terms of total suction. The total suction values, in the region where an abrupt change occurred in these curves, were obtained mainly from these tensiometers. As temperature gradients, monitored by thermocouples installed besides the mini-tensiometers, some 2 cm apart, were typically of the order of 15°C, it is possible that positive pore pressures may have also developed inside the boxes. Thus, the results obtained with the mini-tensiometers, in the range of large volume contraction, may not represent, as a whole, correct values of suction.

Some data values got from the osmotic desiccators, the filter paper not in contact with the residue and the IC tensiometer for the NM soil were compared in order to verify what these techniques were in fact measuring (if total or matric suction). It was observed that the data from the filter paper not in contact adjusted well to those from the desiccators, and both were considered as total suction. The data obtained from the IC probe and the filter paper not in contact were different from each other at high suction levels. But, in the air entry value region, they adjusted rather well. Comparing only the data got from the IC tensiometer and the filter paper in contact with the soil, it was possible to see a good concordance between them. Therefore, it can be considered that the IC probe provided, basically, measures of matric suction. The possibility of its porous stone presenting some osmotic efficiency has, however to be verified.

Figure 11 shows the variation of osmotic suction, obtained by subtracting matric suction from the total one. It can be noted values of osmotic suction around 2 MPa for both residues at 90% of saturation, achieving values up to 9 MPa as the drying process advanced.

Comparing the response of both residues shown in Figures 10 and 11, large differences between them, which can be attributed to the differences in their pore fluid, are observed for volumetric

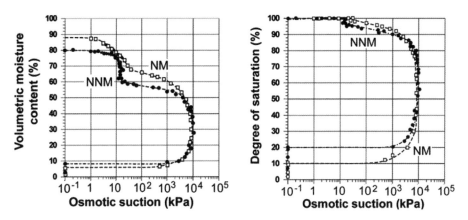

Figure 11. Estimative of osmotic suction variation in the residues.

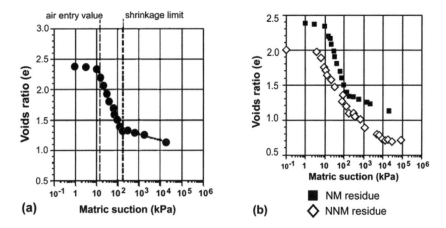

Figure 12. Matric suction–void ratio relationships (a) inflection points (b) comparison of residues.

moisture contents above 50%, when the degree of saturation is still high (above 88%), and for degrees of saturation bellow 30%.

Using the treated results provided by the drying curves, it was possible to obtain the relationships between suction and void ratio shown in Figure 12.

Two inflection points in this correlation, which is equivalent to compressibility curves related to saturated soils, were observed. The first occurred for a void ratio corresponding to the air entry value pressure. The second occurred around a void ratio corresponding to the shrinkage limit. While the matric suction values at the desaturation points of the two residues is not much different, it can be verified a large difference between the matric suction levels corresponding to the shrinkage limit of each residue (order of 200 kPa for the NM residue and of 10 Mpa for the NNM one). Such large difference may be attributed to variations in the structure of the residues resulting from differences in their pore fluid. This, however, has still to be better investigated, as it is not clear whether the initial solids content of these materials affect or not their behaviour when submitted to drying processes.

6 FINAL REMARKS

This paper outlines the methodology used to obtain soil moisture characteristic curves and to evaluate the behaviour of highly compressible materials submitted to a drying process. Two residues from the alumina industry, know as *red muds*, were investigated. They were originated from the same raw material, but presented differing pore fluids. One comprised a caustic pore fluid and, the other, a mixture of this caustic fluid with sulphuric acid.

Four different techniques were used to obtain the relationship between moisture and suction. They were: the filter paper method, small tensiometers installed directly in contact with the wastes inside drying boxes, a suction probe and osmotic desiccators. The knowledge of the osmotic efficiency of tensiometers devices showed to be important when dealing with soils that have pore fluid with different ions concentration than that of the measurement system. The shrinkage process may influence the measurements depending on the technique that is being used.

The samples volume change resulting from suction increase were either measured or evaluated. For such evaluation, a number of relationships among physical indexes obtained during the samples drying process were required. Apart from providing means to define moisture retention relationships, these correlations became a useful tool to infer compressibility parameters of the residues under drying.

From the results presented, a number of aspects are considered relevant and merits further investigation or discussion. For instances, it was shown that the technique employed in characterization tests affects the obtained response, particularly in what refers to grain size distribution and consistency limits. Quite clearly, the use of grains size distribution curves to estimate the SMCC of the residues has to be seen with great care. Also, the shrinkage limit seems to be possible to be identified only from drying tests.

Linear relationships between tensile strength (logarithm scale) and moisture content (natural scale) were obtained. It is not clear whether the maximum tensile strength value corresponds to the moisture content related to the plastic or shrinkage limit, nor if it decreases for moisture contents bellow such limits. As a response to that may be dependent on the cracking pattern developed in the drying process, effects of sample size in the results of such tests have to be investigated.

Differences in drying and moisture characteristic curves of the two residues were observed. As in the drying tests they had different initial solids content, and the results of such tests were employed to obtain suction relationships, further investigation is required to conclude if and when the effect of pore fluid is more relevant than that of the solids content of the slurry submitted to dry.

ACKNOWLEDGEMENT

The authors acknowledge ALCAN, at Ouro Preto, Minas Gerais, Brazil, for the supply of the materials used in the present investigations and to the PRONEX/CNPq project, under way at PUC-Rio, for the support given in the development of the present research.

REFERENCES

ASTM D 5298-92. Standard test method for measurement of soil potential (suction) using filter paper. *Annual book of ASTM Standards*, vol 15.09. pp. 264–268.

Barbour, S. L. 1998. 19th Can. Geotech. Colloq.: The soil-water characteristic curve: a historical perspective. *Can. Geotech. J.* 35, 873–894.

BRITISH STANDARDS INSTITUTE – *Methods of test for soils with civil engineering purposes* – BS 1377 – 1990, HMSO Stat., London.

Bronswijk, J. J. B. 1988. Modelling of water balance, cracking and subsidence of clay soils. *Journal of Hydrology*, 97, 199–212.

Bulkai, D. 1983. *World Review on Environmental Aspects and Protection in Bauxite-Alumina Industry.* Informal Publication.

De Campos, T. M. P., Villar, L. F. S. & Costa Filho, L. M. 1998. Field monitoring of fine waste disposal ponds. Environmental Geotechnics, Seco e Pinto ed. A.A. Balkema. 3rd Int. Congress on Environmental Geotechnics, Lisboa, Portugal, V 1, 253–258.

Favaretti, M. 1995. Tensile strength of compacted clays. *Proc. First Int. Conf. on Unsaturated Soils, UNSAT95.* Paris, pp. 51–56.

Fredlung, D. G. & Xing, A. 1994. Equations for the soil-water characteristic curve. *Can. Geotech. Journ.*, 31, 521–532.

Houston, S. L., Houston, W. N. & Wagner, A. M. 1994. Laboratory filter paper suction measurements. *GTJODJ, ASTM*. Vol 17(2), 185–194.

Ignatius, S. G. & Pinto, C. S. 1991. Aspects on the behaviour of a soil in presence of a caustic chemical effluent. *II REGEO*, R. J., pp. 233–241. (in portuguese)

Jucá, J. F. T. 1990. Behaviour of unsaturated soils under controlled suction. Ph.D. Thesis, Universidad Politécnica de Madrid (in spainish).

Krishnayya, A. V. G. & Einsenstein, Z. 1974. Brazilian tensile test for soils. *Canadian Geotech. Journal*, 11, pp. 632–642.

Lavrov, A. & Vervoort, A. 2002. Theoretical treatment of tangential loading effects on the Brazilian test stress distribution-Technical Note – *International Journal of Rock Mechanics & Mining Sciences*.

Li, L. Y. 2001. A study of iron mineral transformation to reduce red mud tailings. *Waste Management*, 21, pp. 525–534.

Li, L. Y. & Rutherford, G. K. 1996. Effect of bauxite properties on settling of red mud. *Intern. J. Mineral Proc.*, 48, Elsevier, pp. 169–182.

Maciel, I. C. Q. 1991. *Microstructural aspects and geomechanical properties of a facoildal gneisse residual soil*. Ms.C. dissertation. DEC-PUC-Rio, RJ, 182 p.

Marinho, F. A. M. 1994. *Shrinkage behavior of some plastic soils*. Ph.D. Thesis, Imperial College, London.

Mitchell, J. K. 1991. Conduction phenomena: from theory to geotechnical practice. *Géotechnique* 41, n.3 – pp. 299–340.

Oliveira Filho, W. L. 1998. *Verification of a desiccation theory for soft soils*. Ph.D. Thesis, Civil, Environmental and Architectural Dep., University of Colorado, Boulder, USA.

Olsen, P. A. & Haugen, L. E. 1998. A new model of the shrinkage characteristic applied to some Norwegian soils. *Geoderma* 83, pp. 67–81.

Ridley, A. M. & Burland, J. B. 1993. A New Instrument for Measurement of Soil Suction. Technical Note. *Géotechnique* 43(2), 321–324.

Somogyi, F. & Gray, D. 1977. Engineering Properties Affecting Disposal of Red Muds. *Conf. On Geotec. Practice for Disposal of Solid Wastes Materials*, ASCE, Michigan, pp. 1–22.

Swarbrick, G. E. 1992. *Transient Unsaturated Consolidation in Desiccating Mine Tailings*. Ph.D. Thesis, Sch. Civ. Eng., Univ. New South Wales, Australia.

Villar, L. F. S. 1990. *Análise do Comportamento de Resíduos de Processamento de Bauxita: Desenvolvimento de Facilidades Experimentais de Campo e de Laboratório*. M.Sc. Dissertation. DEC Puc-Rio, 252 pp. (in portuguese).

Villar, L. F. S. 2002. *Estudo do Adensamento e Ressecamento de Resíduos de Mineração e Processamento de Bauxita*. Ph.D. Thesis. DEC Puc-Rio, 511 pp. (in portuguese).

Villar, L. F. S., de Campos, T. M. & Vargas Jr, E. 1997. Alguns Aspectos do Fissuramento por Ressecamento de um Resíduo Depositado sob a Forma de Lama. *3° Simp. Bras. Solos Não Saturados – NSAT'97* – vol 2, Rio de Janeiro, RJ, pp. 567–580 (in portuguese).

Villar, L. F. S. & de Campos, T. M. 1999. Ensaios de Ressecamento em Lamas Vermelhas: Estudo de Viabilidade para Uso da Técnica de Dry Stacking. *IV REGEO*, São José dos Campos, São Paulo, Brazil (in portuguese).

Villar, L. F. S. & de Campos, T. M. P. 2001. Obtenção de uma curva característica de sucção pelo uso combinado de técnicas diversas. *4th Brazilian Symp. on Unsaturated Soils, NSAT 2001*- Porto Alegre, RS, Brazil: Gehling & Schnaid Ed. pp 337–353 (in portuguese).

Villar, L. F. S. & de Campos, T. M. P. 2002. Obtaining the soil moisture characteristic curve of a very compressible waste. *Proc. UNSAT2002*. Recife, Brazil, 339–345.

Villar, L. F. S. & de Campos, T. M. P. 2003. *Caracterização geotécnica de resíduos de mineração e processamento de bauxita*. V REGEO, Porto Alegre, Brazil (in portuguese).

Warrick, A. W. et al. 1998. Diurnal fluctuations of tensiometric readings due to surface temperature changes. *Water Resoruces Research*, vol 34 (11), pp. 2863–2869.

Options for modelling hydraulic hysteresis

Y.K. Kazimoglu, J.R. McDougall and I.C. Pyrah
Napier University, Edinburgh, UK

ABSTRACT: Four models of hydraulic hysteresis have been reviewed and compared with experimental data for a compacted sandy silt. For the given primary wetting path, none of the models showed a clear advantage in terms of predictive accuracy. Simplicity of formulation and parameter requirements are thus key factors in the choice of model. In this regard, the scaling-down model appears to be superior to the others.

1 INTRODUCTION

For a given soil, moisture retention is directly related to the soil moisture potential (or suction), and the combined influence of infiltration, redistribution, evaporation and drainage, i.e. the moisture history. When the moisture history includes cycle(s) of drying and wetting, moisture retention is observed to be hysteretic. That is to say, a range of moisture contents may occur at a given soil suction; in other words, the moisture content:suction relationship is non-unique.

The hysteresis phenomenon has been well documented in soil science, hydrology, and irrigation engineering. It has been shown to be significant in controlling groundwater flow in soils and its omission can result in considerable errors in predicted moisture content and pressure head profiles (Kool and Parker, 1987; Stauffer and Kinzelbach, 2001). However, detailed studies of hysteresis are limited to soils with a low range of matric suction, such as Dune sand (Gillham et al., 1976) and Rubicon sandy loam (Topp, 1971). Notwithstanding recent work by Karube and Kawai (2001) and Wheeler et al. (2003), who discussed the influence of hysteresis in the stress–strain behaviour of unsaturated soils, hysteresis in the moisture retention properties of 'engineering' soils has received much less attention.

The extension of a soil model, either flow or stress–strain, to include hydraulic hysteresis must address two issues. The first concerns the representation of the hysteretic moisture content-matric suction relation. The second is the implementation of hysteresis in an efficient manner in the numerical formulation (Gillham, et al., 1976; Mualem, 1984).

Two main types of hysteresis model are distinguished in the literature based on their theoretical foundation; they are empirical and conceptual models. Empirical models are derived from observed moisture retention properties (Dane and Wierenga, 1975; Hanks et al., 1969; Hoa et al., 1977; Scott et al., 1983). Conceptual models are based on independent and dependent domain theories of hysteresis (Poulovassilis, 1962; Mualem, 1974 & 1984). In this paper, the performance of four hysteresis models is examined through the modelling of the hydraulic behaviour of a compacted sandy silt (chosen to provide a range of suctions more relevant to an engineering soil) subjected to a cycle of drying-wetting-drying in the laboratory.

2 MAIN HYSTERESIS LOOP AND SCANNING CURVES

Hysteresis in the moisture content:suction ($\theta:\psi$) relationship is shown in Figure 1. If a fully saturated soil drains, under increasing suction, the $\theta:\psi$ path followed is the main drying curve (MDC). Eventually, moisture content reaches an irreducible value referred to as the residual moisture

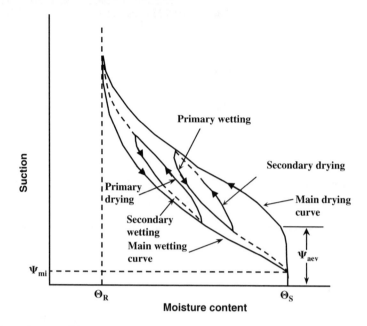

Figure 1. General nature of the hysteresis curves.

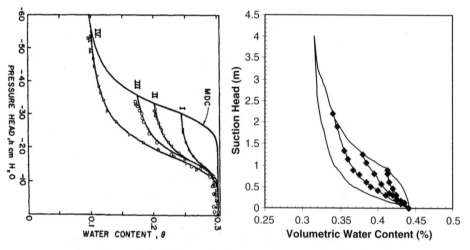

Figure 2. Moisture content-pressure head relation for primary wetting scanning curves (a) of Dune sand measured by Gillham et al., 1976, and (b) Caribou silt Loam measured by Topp (1971).

content θ_r. When the $\theta:\psi$ relationship is hysteretic, wetting from residual moisture content follows a different path, known as the main wetting curve (MWC), which terminates at the saturated moisture content, θ_s. In this simplified description the drying wetting cycle starts and finishes at the same value of θ_s. It is, however, recognised that a common θ_s is only obtained after an initial drying and wetting cycle has been followed, and in the absence of volumetric strains.

The main drying and main wetting curves form the main hysteresis loop, within which lie sets of scanning curves. Consider first reversals in wetting or drying from moisture content:suction combinations lying on the main hysteresis loop but some distance from the residual or saturated moisture contents. Reversals of this kind are referred to as primary scanning curves. For example, reversals

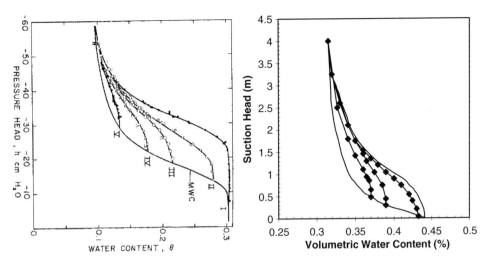

Figure 3. Moisture content-pressure head relation for primary drying scanning curves (a) of Dune sand measured by Gillham et al., 1976, and (b) Caribou silt Loam measured by Topp (1971).

in the form of wetting from the MDC proceed along a primary wetting curve (PWC). Similarly, reversal in the form of drying from the MWC will proceed along a primary drying curve (PDC).

Next, if a reversal occurs while following a primary wetting curve, the moisture content proceeds along a secondary drying curve (SDC). Similarly, if a reversal occurs while following a primary drying curve the moisture content proceeds along a secondary wetting curve (SWC). Higher order curves can be similarly defined.

Figures 2 (a) and (b) present some experimental observations for primary wetting curves measured by Gillham et al. (1976) and Topp (1971), respectively. They show the way in which primary wetting curves converge on θ_s. Similarly, Figure 3 (a) and (b) shows how primary drying curves converge on a residual moisture content θ_r.

3 DESCRIPTION OF MAIN CURVES

A number of expressions defining the $\theta:\psi$ relationship exist. Popular amongst these is that proposed by van Genuchten (1980),

$$\theta = \theta_r + (\theta_s - \theta_r)\left[\frac{1}{1+(\alpha\psi)^n}\right]^m \tag{1}$$

where α and n are curve shape parameters, and m is another shape parameter but often determined from $m = 1 - 1/n$ (Mualem, 1976). The van Genuchten expression can be used to define both MDC and MWC, in which case (given that θ_s and θ_r are common to both MDC and MWC) the main hysteretic loop is defined by six parameters, $\theta_s, \theta_r, \alpha_d, n_d, \alpha_w, n_w$, where subscripts d and w refer to drying and wetting respectively.

4 MODELLING HYSTERESIS: SCANNING CURVES

4.1 Introduction

From a review of hydraulic hysteresis models, four have been selected for further investigation. Selection was made on the ease of parameter identification and reported accuracy of scanning curve interpretation. The first model was proposed by Mualem (1974), and is referred to as Mualem Model II. It differs from the next three in that its derivation is based upon conceptual models of capillary hysteresis. The other three models are empirical: a linear model presented by Hanks et al. (1969), the scaling-down model presented by Scott et al. (1983) and a log-linear model presented by Wheeler et al. (2003). All the models, except that presented by Wheeler et al., define MDC and MWC by the van Genuchten expression (1980).

4.2 Conceptual models

Conceptual models usually refer to the independent domain concept of capillary hysteresis (Neel, 1943). The first application of the independent domain theory to describe moisture retention characteristics was presented by Poulovassilis (1962). According to this theory, a porous medium is a system of independent pore domains, each of which is characterised by a body radius and an opening radius. These two pore radii trigger pore saturation and drainage according to whether the medium is wetting or drying and in so doing produce a hysteresis in the moisture retention relationship. Knowledge of the pore volume distribution function of these pore domains allows determination of the water-filled pore volume and hence moisture content, after any series of wetting and drying.

Note that the conceptual basis enables significant analytical developments to be made; for example, Mualem's Model II has been extended to allow the complete hysteresis loop to be derived from only one branch. The implementation is not, however, trivial so in this paper, the model is based on experimental determination of both branches of the hysteresis loop, which is identical to the main hysteresis loop requirements of the empirical models.

4.2.1 Mualem model II

In Mualem model II, the primary scanning curves are defined as follows:

Primary drying curve,

$$\theta_{1d}^{\psi} = \theta_w(\psi) + [\theta_w(\psi_1) - \theta_w(\psi)]\left(\frac{\theta_d(\psi) - \theta_w(\psi)}{\theta_s - \theta_w(\psi)}\right) \tag{2}$$

Primary wetting curve,

$$\theta_{1w}^{\psi} = \theta_w(\psi) + [\theta_s - \theta_w(\psi)]\left(\frac{\theta_d(\psi_1) - \theta_w(\psi_1)}{\theta_s - \theta_w(\psi_1)}\right) \tag{3}$$

where the volumetric moisture content θ, referenced by subscripts: $1d$ for PDC and $1w$ for PWC, each of which is defined at either ψ, the current suction or ψ_1, the suction reversal point from a main curve to a primary scanning curve. These curves tend smoothly from the reversal point to either residual or saturated moisture contents.

4.3 Empirical models

The procedure used in this approach is to represent the scanning curves by empirical equations based on the shape and properties of the MDC and MWC.

4.3.1 Scaling-down model

Scott et al. (1983) introduced a hysteresis model in which scanning curves are defined by a scaled main curve that passes through the reversal point. For example, a PWC is defined by scaling a MWC to pass through θ_s and the last reversal point $\theta_\Delta : \psi_\Delta (=\psi_1$ in Mualem notation), from drying to wetting. However, if a reversal point is treated as the θ_r in the scaled-down function, the PWC moisture content at the reversal soil suction exceeds the MDC moisture content at the same soil

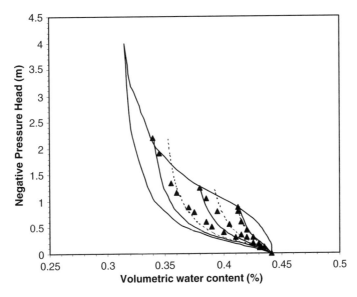

Figure 4. A family of primary wetting curve of Caribou Silt Loam with main hysteresis loop, measured curves (solid symbols), predicted curves using θ_r^* (solid lines) and θ_Δ (dashed lines) for θ_r using van Genuchten's (1980) function.

suction because residual moisture contents are defined at relatively high suctions. An adjustment is therefore required; Scott et al., proposed the following,

$$\theta_r^* = \frac{\theta_\Delta \ (\theta_s - \theta_r) - \theta_s \ (\theta_w(\psi_\Delta) - \theta_r)}{\theta_s - \theta_w(\psi_\Delta)} \quad (4)$$

where θ_r^* is the adjusted residual moisture content for the PWC. The problem is illustrated in Figure 4, which shows a family of measured (solid symbols) PWCs for Caribou Silt (Topp, 1971). Scaled PWCs with θ_r based on reversal points θ_Δ (dashed lines) clearly exceed measured moisture contents particularly at low moisture contents; curves based on adjusted residual values θ_r^* (solid lines) show a much improved fit.

Similarly, drying scanning curves are defined by scaling the MDC to pass through the θ_r and the last reversal point from wetting to drying with an adjusted saturated moisture content,

$$\theta_s^* = \frac{\theta_\Delta \ (\theta_s - \theta_r) - \theta_r \ (\theta_s - \theta_d(\psi_\Delta))}{\theta_d(\psi_\Delta) - \theta_r} \quad (5)$$

These curves tend smoothly from the reversal point to either residual or saturated moisture contents.

4.3.2 Hanks linear model

In the Hanks linear model (Hanks et al., 1969), the primary scanning curves are assumed to be straight lines between MWC and MDC. PDCs are given by,

$$\theta_{1d}(\psi) = \theta_w(h_\Delta) - \left(\frac{\theta_w(h_\Delta) - \theta_d(h_\Delta - \alpha_1\alpha_3)}{\alpha_1\alpha_3}\right)|h - h_1| \quad (6)$$

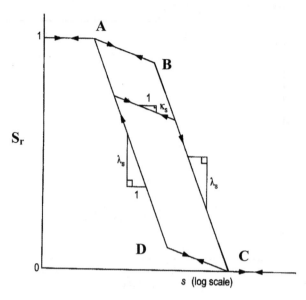

Figure 5. Log-linear model of Wheeler et al. (2003) for hysteretic moisture retention behaviour.

and PWCs by,

$$\theta_w(\psi) = \theta_d(h_\Delta) + \left(\frac{\theta_w(h_\Delta + \alpha_2\alpha_3) - \theta_d(h_\Delta)}{\alpha_2\alpha_3} |h - h_1| \right) \quad (7)$$

where h is the soil-water pressure head, α_1 and α_2 are empirical shape parameters related to the slope of the scanning curves. Parameter α_3 is the maximum difference in suction between the two main curves at any moisture content. Secondary and higher order scanning curves retrace the primary curves until a main curve is reached. PDCs are assumed to have a steeper slope than the PWCs.

With this model, movement along a scanning curve must be continuously monitored and transfer to a main curve instigated to keep moisture content:suction states on or inside the main hysteresis loop.

4.3.3 Log-linear model

The log-linear model, presented by Wheeler et al. (2003), differs significantly from the three models just described. The main hysteresis loop is not defined, rather incremental changes in degree of saturation are defined as either elastic or elasto-plastic processes, according to the current 'modified suction' state (= suction × porosity) in relation to known suction yield conditions. Such a formulation clearly shows the elasto-plastic mechanical origins of the log-linear model.

The form of the moisture retention behaviour is illustrated in Figure 5, from Wheeler et al. (2003). The two main hysteresis curves are represented by log-linear saturation:suction curves of identical slope, λ_s. If a saturation:suction combination lies within the main hysteresis loop, depicted by the parallelogram ABCD, changes in degree of saturation are elastic processes and controlled by parameters κ_s,

$$dS_r = \frac{-\kappa_s\, ds^*}{s^*} \quad (8)$$

Table 1. Summary of parameter and other computational requirements of hysteresis models for primary scanning curves.

Model	Origin	Main curves	Scanning curves	Stored conditions
Mualem model II	Soil science	$\theta_s, \theta_r, \alpha_w, n_w, \alpha_d, n_d$ (van Genuchten)	–	ψ_1
Scaling-down model	Soil science	$\theta_s, \theta_r, \alpha_w, n_w, \alpha_d, n_d$ (van Genuchten)	–	θ_Δ and ψ_Δ
Linear Hanks et al.	Soil science	$\theta_s, \theta_r, \alpha_w, n_w, \alpha_d, n_d$ (van Genuchten)	$\alpha_1, \alpha_2, \alpha_3$	ψ_Δ
Log-linear Wheeler et al.	Geomechanics	λ_s	k_s	s_D^*, s_I^*

where S_r is the degree of saturation and s^* is the modified suction. All elastic saturation:suction curves are assumed to have the same slope; Wheeler et al., acknowledge that this is a relatively crude interpretation.

If a saturation:suction state reaches and then moves along either of the main curves, it is regarded as having 'yielded'. Changes in the degree of saturation are now elasto-plastic processes, of much greater magnitude than that occurring under elastic conditions, and controlled by λ_s,

$$dS_r = -(\lambda_s - \kappa_s)\frac{ds^*}{s^*} \qquad (9)$$

Clearly, hysteresis in the log-linear model requires two material parameters κ_s & λ_s and knowledge of the current suction increase (s_I^*) and suction decrease (s_D^*) yield conditions. As with the Hanks model, movement along a scanning curve must be continuously monitored and transfer to a main curve instigated to keep saturation:suction states on or inside the main hysteresis loop.

4.4 Summary of parameter requirements

- The parameter requirements of the four models under consideration are presented in Table 1. Each of the soil science based models fully defines the main hysteresis loop, which for the van Genuchten formulation requires six parameters.
- Beyond the main hysteresis loop, the Mualem II model has the smallest computational requirement, needing only to store the reversal suction.
- In the scaling-down model reversal suction and moisture content must be stored.
- The Hanks' linear model has a significant parameter requirement in the form of $\alpha_1, \alpha_2, \alpha_3$, not only in the treatment and handling of these parameters but also in their identification.
- Finally, the log-linear model. This model has very few parameter requirements but is a relatively crude representation of the full saturation:suction range. It should be noted that identification of the main and scanning curve parameters, κ_s & λ_s, requires highly specialised laboratory equipment but if such equipment is available, this it may be an attractive option, especially as the measurement of a complete hysteresis loop is not without difficulties, as explained in Section 5.

5 LABORATORY TEST PROGRAMME – INITIAL CONSIDERATIONS

In soil science, experimental moisture retention data including a complete hysteresis loop and scanning curves for wetting and drying processes are usually obtained from coarse-grained soils in which residual moisture contents are reached at relatively low matric suctions. There are fewer

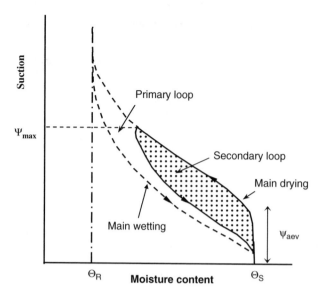

Figure 6. Hysteretic illustrations of the main loop of hysteresis and a secondary loop.

data relating to fine-grained soils. We therefore decided to measure in the laboratory, the hysteretic moisture retention properties of a relatively fine-grained soil, of the kind that might be encountered in engineering practice. There are, however, major practical limitations on the determination of the main hysteresis loop in a fine-grained soil, i.e. a soil in which residual moisture content is only reached at relatively high suction values. If residual moisture contents occur at suction values in excess of 1500 kPa, then the complete hysteresis loop cannot be measured accurately by any one suction measuring technique. In our laboratory, availability of pressure plate ceramics limited the maximum measurable suction to about 500 kPa. Moreover, even at suctions of less than 1500 kPa, where one suction measuring technique is possible, a considerable period of time is required for a fine-grained soil to equilibrate with applied suctions.

It is therefore important to recognise that a fine-grained soil subjected to a drying path, in say a pressure plate extractor, may not reach residual moisture content. Commencement of the wetting cycle at say 500 kPa corresponds to a reversal from MDC and subsequent wetting follows a PWC, as illustrated in Figure 6. Under these circumstances, θ_r, which is a key piece of information in all hysteretic models, is missing.

For the purposes of this test programme we have prepared a soil that is finer-grained than those commonly reported in soil science literature yet possesses a residual moisture content that is reached at applied suctions of around 300 kPa. Thus the complete hysteresis loop can be measured in a pressure plate apparatus.

6 LABORATORY TEST PROGRAMME – TEST METHODS AND MATERIALS

The soil used in this study is compacted sandy silt, which has a relatively low matric suction range compared to fine-grained soils of high clay content. Table 2 summarises some properties of the soil. A conventional pressure plate extractor was modified to measure moisture volume changes of a sample of 90 mm in diameter and 35 mm in height. The moisture volume change in the specimen is measured continuously using a burette that is connected, via an air flushing system, to the ceramic plate inside the pressure chamber. Full details of the experimental set up can be found in Kazimoglu et al. (2003). Suction equilibrium in the specimen was assumed to have been reached when the change in moisture volume was less than 0.2 ml/day, which corresponds to a change of 0.0009 cm^3/cm^3 in the sample.

Table 2. Properties of the test soil.

Specific gravity (Mg/m³)	2.68
Dry unit weight (kN/m³)	17.25
Void ratio	0.52
Sand content (%)	0.43
Fines content (%)	0.57

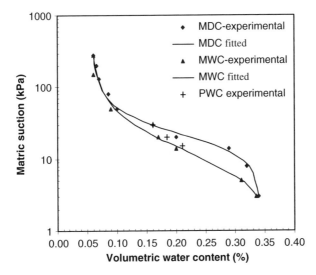

Figure 7. Observed and fitted main curves and observed primary wetting curve for compacted sandy silt.

Table 3. van Genuchten soil moisture parameters.

	MDC	MWC
θ_r	0.060	0.060
θ_s	0.340	0.340
α	0.055	0.10
n	2.80	2.20
m	0.643	0.545

A series of suction increases were applied to an initially saturated sample to measure the MDC. Upon reaching the θ_r, applied suction was decreased. The resulting MDC and MWC data, shown in Figure 7, were collected over a period of four months. van Genuchten parameters fitted to experimental data for the main curves are given in Table 3.

The sample was dried to 30 kPa suction, then re-wetted to 15 kPa. The drying process, beginning from a saturated state, follows MDC, but the reversal at 30 kPa is well below θ_r so subsequent wetting follows a PWC.

7 DEPICTION OF PWC BY HYSTERESIS MODELS

In Figure 8 we show the PWCs predicted by each of the four hysteresis models and the experimental data. Also shown in Figure 8 are two other PWCs, to elaborate the form of the different models at a range of reversal suctions (actually 20 and 15 kPa).

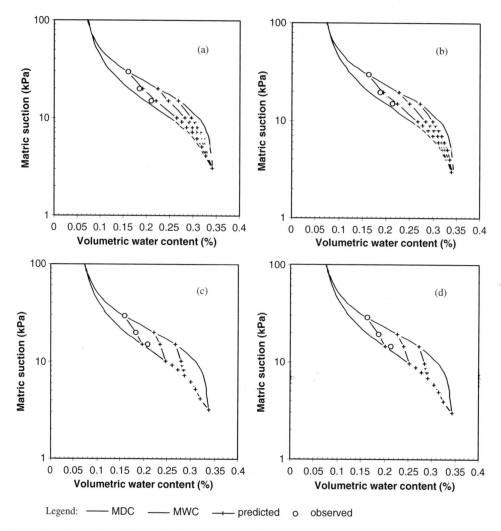

Figure 8. Presentation of the prediction of the six models to the data measured for the compacted sandy silt sample. Solid lines represent the main drying and wetting curves. Symbols were measured data for PWC. Dashed lines and cross symbol represents the predicted curves and points by (a) Model II, (b) Scaling-down, (c) Linear and (d) Log-linear model.

Evidently, for the PWC in question it is difficult to distinguish between the four models. Both Mualem II and the scaling down models tend to overestimate moisture content over the entire PWC. In contrast, the Hanks linear model matches closely the experimental data over the initial portion of the PWC, then tends to underestimate the moisture content as the MWC is approached. This is perhaps not surprising, as this model cannot account for the increasing rate of change of moisture content with decreasing suction shown in Figure 2.

Note that scanning curves in the log-linear model require κ_s to be defined, which cannot be done from main curve data. We have therefore calculated a value of κ_s from the slope of the PWC predicted by linear model, which gives 0.0024 increase in θ with 1 kPa decrease of matric suction, i.e.

$$d\theta = 0.0024 \, d\psi \tag{10}$$

Recognising that in an incompressible medium $d\theta = \theta_s dS_r$ and $ds^* = \theta_s d\psi$, by substituting into equation (10),

$$\kappa_s = \frac{0.0024 \times s^*}{\theta_s^2} = 0.2 \qquad (11)$$

In Wheeler et al., 2003, a trial simulation quotes a value for κ_s of 0.02.

Since comparison of the hysteresis models with this experimental data set does not reveal a clear advantage in terms of accuracy for any one hysteresis model, we conclude our evaluation of the models on implementation issues. For example, it may be noted that:

- Model II uses different expressions for each scanning curve, all of which remain within main hysteresis loop.
- Scaling-down model, while retaining the same level of accuracy of Model II, uses only one form of equation, the van Genuchten form, for all curves. This has significant implications for the numerical implementation since it is necessary to differentiate moisture content with respect to suction to calculate the specific water capacity. Scanning curves remain within the main hysteresis loop.
- Hanks linear model. Appears to be the most parametrically and computationally demanding of the hysteresis models; lacks the similitude of the Mualem model and scaling-down model. Moisture content:suction state must be continuously monitored to ensure scanning paths remain in or on the main hysteresis loop.
- Log-linear model. A relatively simple formulation which is capable of accurately depicting scanning PWC behaviour over a limited range of suctions. Parameters, although few are not easy to define.

8 CONCLUDING REMARKS

The operation and performance of four hysteresis models has been described and tested through their ability to predict primary wetting curve behaviour. The four models were chosen because they had been used with good results in earlier studies. Compared with experimental data for a sandy-silt subjected to a wetting reversal at 30 kPa suction, none of the hysteresis models showed a clear advantage in terms of predictive accuracy. On the data gathered herein, it can be concluded that formulation, parameter identification and implementation issues will control the choice of hysteresis model.

REFERENCES

Dane, J. H., and Wierenga, P. J., 1975. Effect of hysteresis on the prediction of infiltration, redistribution and drainage of water in layer soil. J. Hydrology., 25: 229–242.

Gillham, R. W., Klute, A., and Heermann, D. F., 1976. Hydraulic properties of a porous medium: measurement and empirical representation. Soil Sci. Soc. Am. J., 40: 203–207.

Hanks, R. J., Klute, A., and Heerman, D. F., 1969. A numeric method for estimating infiltration, redistribution, drainage and evaporation of water from soil. Water Resources Research, 5: 1064–1069.

Hoa, N. T., Gaundu, R., and Thirrot, C., 1977. Influence of hysteresis effect on transient flows in saturated-unsaturated porous media. Water Resources Research, 13: 992–996.

Karube, D., and Kawai, K., 2001. The role of pore water in the mechanical behaviour of unsaturated soils. Geotechnical and Geological Engineering, 19: 211–241.

Kazimoglu, Y. K., McDougall, J. R., and Pyrah, I. C., 2003. Moisture retention curve in landfilled waste., (in press) Proc. Intl. Conf. on Unsaturated Soils, Weimar.

Kool, J. B., and Parker, J. C., 1987. Development and evaluation of closed-form expressions for hysteretic soil hydraulic properties. Water Resources Research, 23: 105–114.

Mualem, Y., 1974. A conceptual model of hysteresis. Water Resources Research, 10: 514–520.

Mualem, Y., 1976. A new model for predicting the hydraulic conductivity of unsaturated porous media. Water Resources Research, 12: 513–522.

Mualem, Y., 1984. A modified dependent-domain theory of hysteresis. Soil Science, 137: 283–291.

Neel, L., 1943. Theories des lois d'aimantation de lord raileigh. Cah. Phys., 12: 1–20, 13: 19–30.

Poulovassilis, A., 1962. Hysteresis of pore water, an application of the concept of independent domains. Soil Science, 93: 405–412.

Scott, P. S., Farquhar, G. J., and Kouwen, N., 1983. Hysteretic effects on net infiltration. Advances in Infiltration. ASAE St. Joseph, MI Publications, 11: 163–170.

Stauffer, F., and Kinzelbach, W., 2001. Cyclic hystertic flow in porous medium column: Model, Experiment, and Simulations. Journal of Hydrology, 240: 264–275.

Topp, G. C., 1971. Soil-water hysteresis: The domain theory extended to pore interaction conditions. Soil Sci. Soc. Am. J., 35: 219–225.

van Genuchten, M. Th., 1980. A closed-form equation for predicting the hydraulic conductivity of unsaturated soils. Soil Sci. Soc. Am. J., 44: 892–898.

Wheeler, S. J., Sharma, R. J., and Buisson, M. S. R., 2003. Coupling of Hydraulic Hysteresis and Stress–Strain Behaviour in Unsaturated Soils. Geotechnique, 53: 41–54.

Modelling suction increase effects on the fabric of a structured soil

A. Koliji, L. Laloui, O. Cuisinier* & L. Vulliet
Soil Mechanics Laboratory, Swiss Federal Institute of Technology Lausanne (EPFL), Switzerland
*Current address: Laboratoire Environnement, Géomecanique et Ouvrage, ENSG-INPL, Nancy, France

ABSTRACT: This paper presents the mathematical modelling of the modification of the pore space geometry of a structured soil subjected to suction increase. Structured soil concepts are first introduced considering different fabric units, such as aggregates and fissures. The numerical modelling of the structural evolution is based on experimental test results in which the evolution of the structure of the samples subjected to different suctions is determined using the mercury intrusion porosimetry technique. From this information, the macro and micropore volume evolutions are determined. The results show that drying produces a reduction in the soil total porosity which mainly corresponds to a reduction of the macropore volume. Associated with this phenomenon, an increase in micropore volume is also observed. The proposed model divides pore size distribution into three pore classes (micropores, macropores and non-affected areas). Using the concept of a suction-influenced domain, the proposed model is able to reproduce the main observed fabric evolution between the saturated and dry states.

1 INTRODUCTION

Natural soils almost always show *nested structures*, starting from the scale of the grains and pores and their networks up to field scale, with structures that are 5 to 8 orders of magnitude larger (expressed in their characteristic lengths). Soil structure properties are of great importance in geotechnical, geoenvironmental and agricultural engineering as they influence many soil characteristics, such as compressibility (Labme 1958), hydraulic conductivity (Tamari 1984) or the soil-water characteristic curve (Brustaert 1968) of both the compacted and natural soils. Consequently, improving the understanding of the structural modifications induced by hydric loads is a key issue.

There is a strong relationship between the definition of soil structure and soil heterogeneity. In the first part of this paper, different concepts of structured soils are discussed in order to clarify the approach adopted here.

1.1 *Structured soil*

Natural soils show different kinds of heterogeneity depending on the scale considered. In classical soil mechanics problems, the level of heterogeneity is usually limited to the differences among the characteristics of the soil layers (while at smaller scales, such as that of the pores, heterogeneity can be defined by the means of smaller structural units, such as different pores, fissures or aggregates). Inside a specific soil layer, at a smaller scale, the soil is usually assumed to be homogeneous as soon as enough grains or aggregates are considered, the pore-scale heterogeneities being smoothed out (Laloui et al. 2003). This implies the consideration of a representative elementary volume (REV) of a given material. A REV is required to include a significant amount of small-scale heterogeneity to respect a meaningful statistical average (Vogel & Roth 1998). As an example, the dependence of the sample porosity on its scale is shown in Figure 1. It can be seen that the REV may be representative of a homogeneous medium, while smaller samples are completely heterogeneous. For volumes below the REV, as may be seen in Figure 1a, the whole sample can be located completely within the pores or within the particles, leading to quite different values of porosity. However, when the

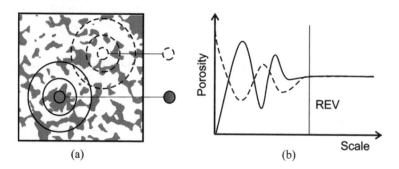

Figure 1. Effect of sample size on the porosity evolution: (a) Sample sizes; (grey = solid, white = void). (b) Porosity vs. sample scale.

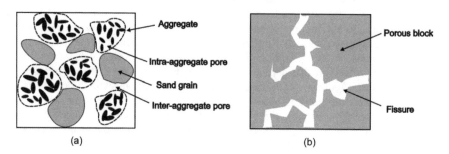

Figure 2. Structured soils: (a) Aggregated soil; (b) Fissured porous medium.

sample sizes increase, the difference between them decreases and the obtained values of porosity tend to a single one (obtained at the REV level).

Natural soils show different structures, characterised by structural units, such as aggregates, porous block, fissures, earth worm holes and root channels (especially in the top layers of agricultural soils). It is widely accepted in the literature (Al-Mukhtar et al. 1996, Cuisinier & Laloui 2004, Delage et al. 1996, Gens et al. 1995, Griffiths & Joshi 1990) that these soils, especially the compacted ones, have two levels of structure: macro and micro. The soil microstructure can be defined as the elementary particle associations within the soil aggregates (Al-Mukhtar et al. 1996), whereas the arrangement of these soil aggregates and the relation among the structural units at the aggregate level is referred to as the macrostructure. Consequently, the pores of a soil consist of two main classes, micropores and macropores, corresponding to the two levels of soil structure. These two different levels play important roles in the hydro-mechanical behaviour and deformation process of soils (Cuisinier & Laloui 2004). For instance, although micropores are usually neglected, they have a strong influence on water retention characteristics. The larger pores (macropores) dominate the dynamics of solute transport. They also play a crucial role in cases of soil freezing and indirectly affect heat transfer.

1.2 Concept of double porosity

Such a division of pores into macro and micro levels leads to the concept of double porosity for soils. This concept can be used for soils with two distinct levels of porosity. Figure 2 shows two different types of structured soil for which the concept of double porosity can be used. In Figure 2a, an aggregated soil is presented. Here, the pores between the aggregates (inter-aggregate) are macropores and the pores within individual aggregates (intra-aggregate) form micropores. Figure 2b shows a fissured porous medium. It may be thought of as a number of porous blocks separated from each other by a system of randomly distributed fissures. Thus, here again, macropores correspond to fissures and micropores to pores of porous blocks (Valliappan & Khalili 1990).

Although usually the single porosity models are seen to be fairly successful to describe the behaviour of porous materials, they are not suitable for structured double porosity (fissured or aggregated) materials. In such systems, although most of the fluid mass is stored in the micropores, the permeability of the macropores is much higher than that of the micropores (Tuncay & Corapcioglu 1995). This leads to a dually permeable medium and also to two distinct fluid pressure fields: one in the macropores and the other in the micropores. Therefore, the fluid pressures contained in the two types of pores may reach equilibrium at different rates, and virtually independently, if the channels for fluid transport between the two types of pores are restricted. This concept leads to models in which distinct pore pressures are present in the matrix and in the fractures (Wang & Berryman 1996).

The double porosity concept was first introduced by Barrenblatt et al. (1960) as a flow model for non-deformable fissured porous media. This model was simplified by Barrenblatt (1963) and Warren & Root (1963). Then Aifantis (1977) extended this concept by sketching the basis of a multi-porosity model to include mass exchange of solutes through diffusion between the mobile and stagnant water phases. The general theory of Aifantis unified the earlier proposed models of Barenblatt for fluid flow through non-deformable porous media with double porosity and Biot's theory for consolidation of deformable porous media with single porosity. Khalili & Valliappan (1996) added a new coupling link between the elastic volumetric deformations of the two pore systems. Khalili et al. (2000) presented a fully coupled thermo-hydro-elastic model for double porous media using a macroscopic approach.

In most of the previous research work, an attempt was made to explain and to model the hydro-mechanical behaviour of double porous media based on general assumptions concerning soil structure evolution during loading. Very few of them considered unsaturated soils.

1.3 Experimental characterisation of the soil structure

From the rapid synthesis presented above, soil structure may be seen as the arrangement and organisation of particles in soil. In order to be able to study this soil structure, quantitative approaches are required. One of them is based on the analysis of the pore size distribution (PSD) of soils, widely accepted to be representative of soil structure (Al-Mukhtar 1995, Delage et al. 1996, Delage & Lefebvre 1984). Pore size distribution of soils (normally called soil fabric) corresponds to the geometrical arrangement of soil particles.

In the case of structured double porosity soils, a few studies have been undertaken to characterise fabric changes induced by various types of hydro-mechanical loads. Delage & Lefebvre (1984), studying compacted Champlain clay fabric with mercury intrusion porosimetry (MIP) and scanning electron microscopy, have shown that, during consolidation, only the largest pores collapse at a given stress increment. Small pores are only compressed when all of the macropores have been completely closed by loading. These results were confirmed by other experimental studies (Griffiths & Joshi 1990, Coulon & Bruand 1989, Lapierre et al. 1990). Soil fabric is also sensitive to suction increase, as evidenced by other authors (Simms & Yanful 2001) who investigated the fabric of a compacted glacial till under different suctions, from saturation up to a matric suction of 2500 kPa. They have shown that suction increase brought about a progressive rise in the microporosity associated with macroporosity reduction. Simms & Yanful (2001) investigated the relationship between suction and induced fabric modification. Their results have demonstrated that the soil fabric is extremely sensitive to any suction variation. They transformed the measured pore size distribution to take account of pore trapping and developed a model for shrinkage of pores due to suction. Using this model, they proposed a method to predict the changes in pore size distribution during drying tests (Simms & Yanful 2001, 2002).

In order to model the suction-induced modification in a structured soil, a mathematical model based on experimental results of suction increase on a compacted soil is presented in this paper. The evolution of the double porosity structure of the soil is determined using mercury intrusion porosimetry. The proposed numerical model is presented and validated in Section 3.

2 EXPERIMENTAL RESULTS

As a part of a comprehensive research program (Cuisinier & Laloui 2004), experimental tests were carried out for the analysis of fabric modifications brought about by suction increase. The tested soil was a sandy loam from the eastern part of Switzerland (morainic soil of the Swiss central plateau). The plasticity index of the soil was approximately $I_p = 12\%$ and its liquid limit close to $w_L = 30\%$. The unit weight of the soil particles was $\gamma_s = 26.1\,\text{kN}\,\text{m}^{-3}$. In this study, all of the samples were prepared using the same procedure. After sampling in the field, the soil was air dried and gently crushed after several days. Aggregates between 0.4 and 2 mm were selected by sieving and were then wetted up to a mass water content of about $w = 14\%$ and stored in an airtight container for at least one week in order to reach moisture equilibrium. The aggregates were then statically compacted directly inside the testing equipment up to a target dry unit density of $\gamma_d = 14\,\text{kN}\,\text{m}^{-3}$. In the last stage, the samples were saturated by imposing a low water head, of a few centimetres, at the base of a Richard cell. After this saturation phase, the mass water content of the samples was between 36 and 37%. The initial state of the samples, after the saturation phase, corresponded to zero suction.

In a pressure plate device (air overpressure method for suction control, (Richards 1935)), the suction was increased by steps from the saturated state to 50, 100, 200 and 400 kPa. At each step, when equilibrium was reached, samples were taken for fabric determination by mercury intrusion porosimetry and for water content determination. The obtained water characteristic curve is presented in Figure 3. The repeatability of the procedure was successfully checked and it was shown that the freeze-drying technique used for the MIP had no influence on the pore size distribution (see Cuisinier & Laloui (2004), for more details).

2.1 Soil fabric determination with mercury intrusion porosimetry (MIP)

Several methods exist to determine soil fabric: scanning electron microscopy, optical microscope, nitrogen adsorption, mercury intrusion porosimetry (MIP), etc. For this study, the MIP technique was selected because it allows the measurement of a wide pore size range, from a few nanometres up to several tens of micrometers, and it permits the identification of the different soil pore classes. The theoretical bases for the determination of soil fabric with MIP are very similar to those of the pressure plate test. In the case of MIP, the non-wetting fluid is mercury and air is the wetting fluid. The mercury pressure is increased by steps and the intruded volume of mercury is monitored for each pressure increment. Assuming that soil pores are cylindrical flow channels, Jurin's equation

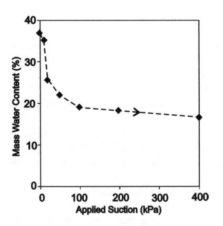

Figure 3. Soil–water characteristic curve of the tested material.

is used to determine the pore radius associated with each mercury pressure increment:

$$R = \frac{2 T_s \cos\alpha}{\psi} \quad (1)$$

where R is the entrance pore radius, T_s the surface tension of the liquid (0.485 N.m^{-1} for mercury), α the contact angle of fluid interface to solid (140° is an average value for mercury-air interface as suggested in Penumadu & Dean (2000), and ψ the pressure difference between the two interfaces (in Pa). From the MIP test, the injected mercury volume can be measured at each step and the cumulative mercury volume injected can be obtained as a function of the equivalent pore radius. One method to present the results of an MIP test is to use the pore size density function which represents the volume of pores for each existing radius as defined by (Juang & Holtz 1986):

$$f(\log d_i) = \frac{\Delta V_i / V_t}{c} \quad (2)$$

$$\log d_i = \frac{1}{2}(\log R_{i-1} + \log R_i) \quad (3)$$

where c is a constant value related to the difference between radius classes (in the present results $c = 0.3$). ΔV_i is the injected mercury volume at a given pressure increment corresponding to a pore class of pores having a radius between R_i and R_{i-1}, V_t the total injected mercury volume and d_i is the midpoint of the pore class which is calculated by Equation 3. The use of a logarithmic scale is due to the wide range of pore radii, i.e. several orders of magnitude. It should be pointed out that the determination of PSD with the MIP test is influenced by several experimental side effects, such as pore entrapment, pore neck, etc. As a consequence, the measured PSD might differ from the real PSD of the tested soil and the pore radius associated with each pressure increment is only an entrance pore radius. As a simplification, the term "entrance" will be omitted in the following when dealing with pore radii calculated from MIP data. More details about these phenomena are available in Delage & Lefebvre (1984).

Each soil sample was carefully cut in small pieces of approximately 1 g. Due to technical requirements, the MIP test must be conducted on totally dried soil pieces. Among available dehydrating methods, the freeze-drying method was considered as the least disturbing preparation technique for water removal (Gillot 1973). Soil pieces were quickly frozen with liquid nitrogen (temperature of −196°C) and then placed in a commercially available freeze-drier for approximately 1 day for the sublimation of the water. The samples were subsequently kept inside desiccators until the MIP tests were performed with a Porosimeter 2000 device (Carlo Erba Instruments).

2.2 Suction increase effect on PSD

The pore size density function (PSD) for five suction levels is shown in Figure 4. The PSD of the sample at suction equal to zero (initial state) had a shape very similar to that of a silt compacted dry of optimum (Delage et al. 1996), as two pore classes can be seen (micro and macropores). The limit between the two pore classes is about 1 μm.

Knowing the sample's initial fabric (zero suction), the analysis of the influence of drying on fabric was carried out. It may be seen in Figure 4a that the suction increase induces modifications of the soil fabric, such as a strong reduction of the macroporosity. It is interesting to note that the reduction of the amount of macropores is accompanied by a relative increase in the microporosity, which can be explained by the shrinkage of the macropores. The comparison of all of these PSD curves shows that the drying process did not affect pores smaller than 0.1 μm.

Another way to present the MIP results is to give the volume fraction of pores per unit weight of soil sample (Fig. 4b). This provides more physically comparable tools and also facilitates the interpretation of results. In Figure 4b, the lower limit of the pore radii corresponding to high pressures has been omitted; this eliminates possible errors due to measurement at high pressures.

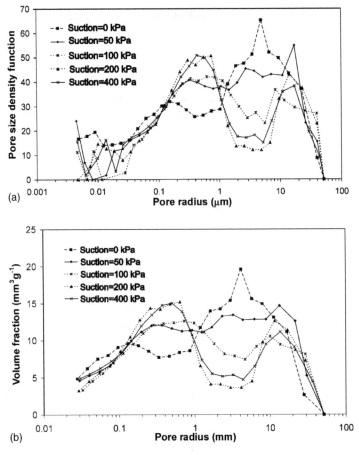

Figure 4. Modification of samples fabric when suction is increased from 0 up to 400 kPa: (a) Pore size density function, (b) Volume fraction.

3 MODELLING PORE SIZE DISTRIBUTION MODIFICATION

In this section, a mathematical model is presented to simulate the PSD modification during the suction increase based on the PSDs of the saturated and the drier cases. Due to the fact that the concept of volume fraction is more suitable for the physical understanding of behaviour, the proposed model was developed on this basis. A validation of the model results with respect to experimental ones then concludes this section.

3.1 Characterisation of pore classes

The modelling concept is as follows: given two limit cases, the saturated case (zero suction – initial case) and the driest case (suction of 400 kPa – final case), prediction of the modification of the PSD curves for intermediate cases.

From the results in Figure 5, four different zones of pore classes can be recognised: Zones 1 and 4 are related to the pores which are only slightly affected by suction and on which the influence of suction can be neglected. Zone 2 corresponds to the micropores; the volume fraction of these pores increases as suction increases. The pore volume fraction of the pores that belong to the third zone reduces as suction increases. The behaviour of these pore classes are different and they should

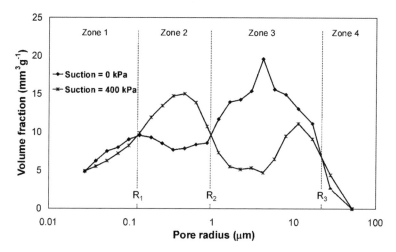

Figure 5. Different zones of pore classes.

be distinguished in the model formulation. Three pore radii R_1, R_2 and R_3 determine the limits between pore zones.

3.2 Suction influence domain

The relationship between any suction ψ and the corresponding pore radius R can be determined by Equation 1 using the value of contact angle α equal to zero (air–water interface):

$$\psi = \frac{2\,T_s}{R(\psi)} \qquad (4)$$

where the surface tension of the water, T_s, is equal to $0.07275\,\text{N m}^{-1}$. $R(\psi)$ is the drained pore radius at that value of suction ψ and it can be considered as the minimum pore radius influenced by suction ψ (note that in doing so, hysteretic effects are neglected).

Using Equation 4, it is be possible to determine the suction influence domain for a given suction value. This domain can be defined as a domain in which the pores have radii greater than $R(\psi)$. Thus, Zone 3 is subdivided into Zones 3a and 3b when $R_2 < R(\psi) < R_3$ (see Figure 6).

3.3 Modelling steps

The following general framework to model the PSD modifications is outlined as follows:

– volume fraction of pores of Zone 1 and 4 are not affected by suction increase;
– as suction increases, only the pore volume of Zone 3 (macropores) can decrease;
– a part of the volume fraction decrease of Zone 3 will be added to Zone 2 and consequently the volume fraction of Zone 2 (micropores) will increase;
– it is assumed that, at a given suction, only the volume fraction of pores of Zone 3 located in the suction influence domain is added to the volume fraction of pores of Zone 2. This concept is illustrated in Figure 6.

Based on this framework, PSD (or volume fraction distribution) modification at a given suction ψ can be evaluated by considering five intervals of radii.

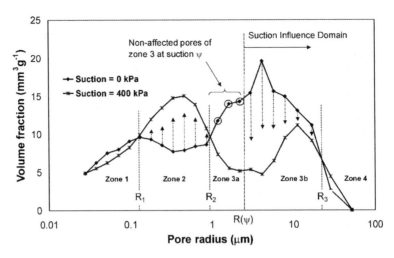

Figure 6. Suction influence domain and evolution of pore classes.

3.3.1 *Zone 1*

As already mentioned, volume fractions of pores located in Zone 1 are only slightly affected by suction. Therefore, in this model, it is assumed that these pores keep their initial value of volume fraction. This implies:

$$v(r,\psi) = v(r,\psi_0) \qquad ; r < R_1 \qquad (5)$$

where r is the pore class radius, v(r, ψ) is the predicted volume fraction of pores at suction ψ and v(r, ψ_0) is the volume fractions of pore class measured at the initial stage (zero suction, $\psi = 0$).

3.3.2 *Zone 4*

Here again, no changes are expected, and thus:

$$v(r,\psi) = v(r,\psi_o) \qquad ; r > R_3 \qquad (6)$$

3.3.3 *Zone 3a*

For the pores located in Zone 3 but not in the suction influence domain (Zone 3a, Fig. 6), no reduction in their volume fraction is experienced and, thus:

$$v(r,\psi) = v(r,\psi_o) \qquad ; R_2 < r < R(\psi) \qquad (7)$$

3.3.4 *Zone 3b*

In this zone, the volume fraction of each pore class can be obtained by subtracting a specific value of volume fraction from the initial PSD. As already assumed for a given suction, only pores of Zone 3 located in the suction influenced domain (Zone 3b) show reduction in their volume fractions. It can be concluded that the total volume fraction that should be subtracted from the initial case (saturated) during the suction increase from the initial to final case is calculated by:

$$V_\psi = \sum_{r > R(\psi)} (v(r,\psi_o) - v(r,\psi_f)) \quad ; (zone\,3, \; R_2 < R(\psi) < R_3) \qquad (8)$$

where V_ψ is the *total influenced volume fraction*, v(r, ψ_0) is the initial volume fraction of pore class with radius r at suction ψ_0, and v (r, ψ_f) is the final volume fraction at suction ψ_f of the final case.

As a result of pore trapping, only a portion of V_ψ should be subtracted from the initial PSD, called the effective volume fraction, $V_{\psi e}$. As a first approximation, the relationship between V_ψ and $V_{\psi e}$ can be assumed as:

$$V_{\psi e} = C_3 \cdot V_\psi \tag{9}$$

where $C_3 = \tilde{C}_3(\psi)$ is a function of ψ (to be defined). Knowing $V_{\psi e}$, the new location of the PSD for a specific suction ψ can be found by linear interpolation between the initial and the final cases:

$$v(r,\psi) = v(r,\psi_0) - \frac{V_{\psi e}}{V_\psi} \cdot (v(r,\psi_0) - v(r,\psi_f)) \quad ; \quad R(\psi) < r < R_3 \tag{10}$$

and by substituting Equation 9 into Equation 10:

$$v(r,\psi) = v(r,\psi_0) - C_3 \cdot (v(r,\psi_0) - v(r,\psi_f)) \quad ; \quad R(\psi) < r < R_3 \tag{11}$$

Due to lack of experimental evidence on pore trapping, an approximation should be made for C_3. A possible expression for $C_3 = \tilde{C}_3(\psi)$ can be constructed by considering that:

- for $1R(\psi) < R_2$, $C_3 = 1$ because the PSD curve obtained by MIP is assumed to be the real curve without air trapping.
- for $R(\psi) = R_3$, $V_\psi = 0$ and C_3 can take any value, $V_{\psi e}$ being always zero.
- for intermediate cases, the trapped air is expected to be a function of the ratio V_ψ/V_3, where V_3 is the total volume fraction difference between the initial and final cases in Zone 3:

$$V_3 = \sum_{R_2 < r < R_3} (v(r,\psi_0) - v(r,\psi_f)) \tag{12}$$

Back-calculations from the authors' results showed that a possible candidate for $C_3 = \tilde{C}_3(\psi)$ is:

$$C_3 = \left(\frac{V_\psi}{V_3}\right)^2 \tag{13}$$

3.3.5 Zone 2

In order to model the volume fraction variation of pores in Zone 2, some information about the amount of affected macropore volume fraction which is going to be added to the micropores is required. Volume variation of a soil during the application of suction can be divided in two parts. The first one is the external volume variation which can result in a total porosity reduction, and the second one is the internal volume variation which is related to the shrinkage of macropores and subsequent change into micropores. Therefore, for each macropore class located in Zone 3, there is only a portion of the total experienced volume fraction reduction which is added to the micropore classes. It is assumed that, at each suction step, this portion is determined by a constant coefficient which is equal to the ratio of total micropore increase to the total macropore decrease from the initial to the final case. Therefore, this coefficient can be determined as follows:

$$C_2 = \frac{\sum_{R_1 < r < R_2} (v(r,\psi_f) - v(r,\psi_0))}{\sum_{R_2 < r < R_3} (v(r,\psi_0) - v(r,\psi_f))} \tag{14}$$

The total moving volume fraction from macropores to micropores, V_{2e} which is a portion of the total influenced volume fraction $V(\psi)$ is determined by this coefficient:

$$V_{2e} = C_2 \cdot V_\psi \tag{15}$$

This volume fraction should be distributed among the micropores and added to the initial volume fractions. The new location of the PSD for a specific suction ψ can be found (as in Equation 10) by linear interpolation. As a result, the modelled volume fraction of pores located in Zone 2 can be estimated by:

$$v(r,\psi) = v(r,\psi_0) + \frac{V_{2c}}{V_2} \cdot (v(r,\psi_f) - v(r,\psi_0)) \quad ; \quad R_1 < r < R_2 \quad (16)$$

where:

$$V_2 = \sum_{R_1 < r < R_2} (v(r,\psi_f) - v(r,\psi_0)) \quad (17)$$

4 NUMERICAL VALIDATION OF THE MODEL

The model presented is conceived for the simulation of the modification of the pore space geometry of double porosity structured soils subjected to a suction increase. It is mainly based on the concept of the suction influence domain and includes experimental evidence of the effect of suction on micropores and macropores. In order to validate this model, it was used to simulate the PSD modification for the experiments presented in Section 2 of this paper.

The initial data needed for the model are the PSD curves for the saturated and drier states (suction of 400 kPa). The model predictions made for the PSD curves at suction levels of 50, 100 and 200 kPa will be analysed. The comparison between the simulated curves with the experimental results is given in the following paragraphs.

Figure 7 shows the comparison of the modelled and measured PSD curves at a suction of 50 kPa. To show the evolution of the PSD curve from the saturated state (zero suction) to a suction of 50 kPa, the initial PDS curve (saturated) was included.

As may be seen in Figure 7, the model correctly reproduces the decrease of macropores as well as the increase of micropores. The lower limit domain of the suction influence is situated at a radius of about 3μm. The modelled macropores under this limit have the same values of volume fraction as those for the saturated case. This aspect was also observed for the experimental points. The same satisfactory prediction applies to the behaviour at a suction of 100 kPa (Fig. 8), even if the experimental and the simulated results are not perfectly superposed. Almost all macropores are affected by the suction increase. The numerical simulation at a suction of 200 kPa is shown in

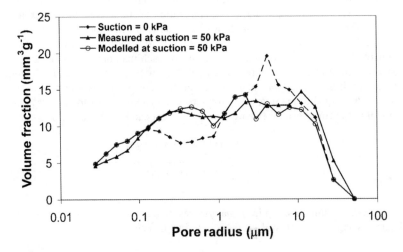

Figure 7. Modelled and measured PSD at suction 50 kPa.

Figure 9. The model reproduces the behaviour of the three zones remarkably well. In Zone 1 (pore radius less then R_1 and greater than R_3) the effect of suction is very limited. In Zone 3 (macropores), the important decrease in the volume fraction is well reproduced. This is also the case for Zone 2 (micropores), where the significant increase in micropores is well predicted. The modelled volume fractions at 200 kPa are the same as those at 400 kPa due to the fact that this latter one is the final limit of the model.

5 CONCLUSIONS

In this paper, a mathematical model is presented for the numerical simulation of the modification of the pore space geometry of a structured double porosity soil subjected to suction increase. The model is mainly based on the concept of the suction influence domain backed up by experimental evidence. The model divides the pore size domain into different zones: macropores, micropores and domains where the suction effect is limited. Each zone has its own behaviour and the micropore

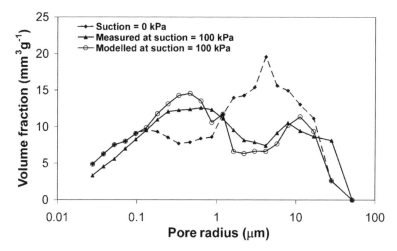

Figure 8. Modelled and measured PSD at suction 100 kPa.

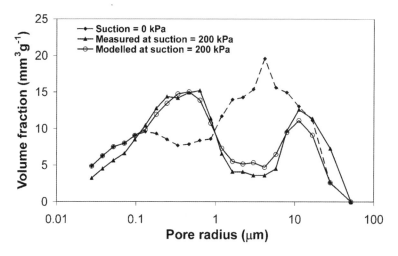

Figure 9. Modelled and measured PSD at suction 200 kPa.

and the macropore zones are strongly coupled (a part of the volume lost in the macropores is gained by the micropores). The numerical simulations presented show that the model is able to reproduce the main aspects of suction induced effects on soil structure.

REFERENCES

Aifantis, E.C. 1977. Introducing a Multi-Porous Medium. *Developments in Mechanics* 8: 209–211

Al-Mukhtar, M., Belanteur, N., Tessier, D. & Vanapalli, S.K. 1996. The fabric of a clay soil under controlled mechanical and hydraulic stress states. *Applied Clay Science* 11: 99–115

Barrenblatt, G.I. 1963. On certain boundary value problems for the equation of seepage of liquid in fissured rock. *Journal of Applied Mathematics and Mechanics : translation of the Soviet journal : Prikladnaja Matematika i Mekhanika (PMM)* 27: 513–518

Barrenblatt, G.I., Zeltov, I.P. & Kochina, N. 1960. Basic concepts in the theory of seepage of homogeneous liquids in fissured rocks. *Journal of Applied Mathematics and Mechanics : translation of the Soviet journal : Prikladnaja Matematika i Mekhanika (PMM)* 24(5): 1286–1303

Brustaert, W. 1968. The permeability of a porous medium determined from certain probability laws for pore-size distribution. *Water Resources Research* 4: 425–434

Coulon, E. & Bruand, A. 1989. Effects of compaction on the pore space geometry in sandy soils. *Soil and Tillage Research* 15: 137–152

Cuisinier, O. & Laloui, L. 2004. Fabric evolution during hydromecanical loading of a compacted silt. *International Journal for Numerical and Analytical Methods in Geomechanics, in press*

Delage, P., Audiguier, M., Cui, Y.-J. & Howat, M.D. 1996. Microstructure of compacted silt. *Canadian Geotechnical Journal* 33: 150–158

Delage, P. & Lefebvre, G. 1984. Study of the Structure of a Sensitive Champlain Clay and of Its Evolution During Consolidation. *Canadian Geotechnical Journal* 21(1): 21–35

Griffiths, F.J. & Joshi, R.C. 1990. Change in pore size distribution due to consolidation of clays. *Géotechnique* 40(2): 303–309

Juang, C.H. & Holtz, R.D. 1986. A Probabilistic Permeability Model and The Pore Size Density Function. *International Journal for Numerical and Analytical Methods in Geomechanics* 10: 543–553

Khalili, N. & Valliappan, S. 1996. Unified theory of flow and deformation in double porous media. *European Journal of Mechanics A/Solids* 15(2): 321–336

Labme, T.W. 1958. The engineering behaviour of compacted clays. *Journal of the Soil Mechanics and Foundation Division ASCE* 84: 1–35

Laloui, L., Klubertanz, G. & Vulliet, L. 2003. Solid-Liqiud-Air Coupling in Multiphase Porous Media. *International Journal for Numerical and Analytical Methods in Geomechanics* 27: 183–206

Lapierre, C., Leroueil, S. & Locat, J. 1990. Mercury intrusion and permeability of Louisville clay. *Canadian Geotechnical Journal* 27: 761–773

Penumadu, D. & Dean, J. 2000. Compressibility effect in evaluating the pore-size distribution of kaolin clay using mercury intrusion porosimetry. *Canadian Geotechnical Journal* 37(2): 393–405

Richards, R.A. 1935. Capillary conduction of liquids through porous medium. *Physics* 1: 318–333

Simms, P.H. & Yanful, E.K. 2001. Measurement and estimation of pore shrinkage and pore distribution in a clayey till during soil-water characteristic curve tests. *Canadian Geotechnical Journal* 38(4): 741–754

Simms, P.H. & Yanful, E.K. 2002. Predicting soil-water characteristic curves of compacted plastic soils from measured pore-size distribution. *Géotechnique* 52(4): 269–278

Tamari, S. 1984. Relation between pore-space and hydraulic properties in compacted beds of silty-loam aggregates. *Soil Technoligy* 7: 57–73

Tuncay, K. & Corapcioglu, M.Y. 1995. Effective stress principle for saturated fractured porous media. *Water Resources Research* 31(12): 3103–3106

Valliappan, S. & Khalili, N. 1990. Flow through fissured porous media with deformable matrix. *International Journal for Neumerical Methods in Engineering* 29: 1079–1094

Vogel, H.J. & Roth, K. 1998. A new approach for determining effective soil hydraulic functions. *European Journal of Soil Science* 49: 547–556

Wang, H.F. & Berryman, J.G. 1996. On Constitutive Equations and effective stress principles for deformable, double-porosity media. *Water Resources Research* 32(12): 3621–3622

Warren, J.R. & Root, P.J. 1963. The behaviour of naturally fractured reservoirs. *Society of Petroleum Engineering Journal*: 245–255

A bounding surface plasticity model for unsaturated clay and sand

A.R. Russell
Department of Civil Engineering, University of Bristol, Bristol, United Kingdom

N. Khalili
School of Civil and Environmental Engineering, University of New South Wales, Sydney, NSW, Australia

ABSTRACT: This paper presents a simple bounding surface plasticity model that is suited to describe the stress–strain behaviour of both unsaturated clays and unsaturated sands. The model is presented in a critical state framework using the concept of effective stress. The versatility of the model is highlighted by its ability to simulate using single sets of suction independent material parameters: (1) the stress–strain behaviour of unsaturated speswhite kaolin subject to three load paths; and (2) the stress–strain behaviour of an unsaturated quartz sand subject to two load paths.

1 INTRODUCTION

In recent years, several constitutive models have been proposed to describe the stress–strain behaviour of unsaturated soils based on the modified Cam-Clay model. Nevertheless, they are specifically suited to fine grained soils such as silts and clays and not to coarse grained soils such as sands and gravels.

In this paper, an elasto-plastic model is presented that is sufficiently general such that it is suited to all types of unsaturated soils. The model is formulated using bounding surface plasticity theory due to its versatility (Dafalias and Popov, 1975; Dafalias, 1986). It is presented in a critical state framework using the concept of effective stress. Both pre- and post-peak plasticity are taken into account. The direction of loading is controlled by a function that is capable of accounting for the varied responses of different soil types subjected to different loading conditions. A non-associated flow rule is adopted and isotropic hardening/softening of the soil matrix results from a change in suction as well as plastic volumetric strain. Consideration is given to the effects of unsaturation in the definition of the critical state. Consideration is also given to the effects of particle crushing in sands at high stresses.

The model is calibrated for speswhite kaolin using the triaxial compression test results for three load paths presented in Wheeler and Sivakumar (1995); and for Kurnell sand, a predominantly quartz sand containing no fines, using the triaxial compression and oedometric compression test results reported in Russell (2004).

2 THE BOUNDING SURFACE PLASTICITY MODEL

The model presented here is an extension of that appearing in Russell and Khalili (2004a), and was initially developed to describe the stress–strain behaviour of saturated Kurnell sand across a range of stresses including those sufficient to cause particle crushing and subjected to varied load paths.

2.1 *Effective stress*

The effective stress (Bishop, 1959) is defined as:

$$\sigma' = \sigma_n + \chi s \qquad (1)$$

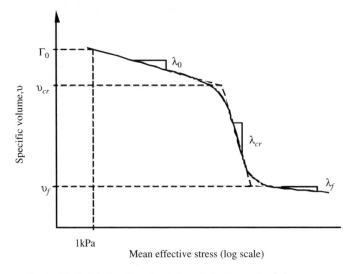

Figure 1. A generalized critical state line for saturated sands in the $\upsilon \sim \ln p'$ plane.

where σ_n is the total stress in excess of pore air pressure (u_a) also referred to as net stress; s is the difference between pore air and pore water pressure ($u_a - u_w$) also referred to as suction; and χ is the effective stress parameter and has a value of 1 for saturated soils and 0 for dry soils.

2.2 Notation

Conventional triaxial $p' - q$ notation is adopted, where p' is the mean effective stress and q is the deviator stress. The corresponding work conjugate strain variables are the soil skeleton volumetric strain ε_p and shear strain ε_q. Compression is assumed positive. The pairs of stresses and strains are abbreviated in vector form $\boldsymbol{\sigma} = [p', q]^T$ and $\boldsymbol{\varepsilon} = [\varepsilon_p, \varepsilon_q]^T$. Incremental elastic strains are linked to the incremental stress invariant through $\dot{\boldsymbol{\sigma}} = \mathbf{D}^e \dot{\boldsymbol{\varepsilon}}^e$ where $\mathbf{D}^e$ is the elastic compliance matrix. The plastic stress–strain relationship is of the form $\dot{\boldsymbol{\varepsilon}}^p = \mathbf{n}^T \mathbf{m} \dot{\boldsymbol{\sigma}} / h$ where $\mathbf{n} = [n_p, n_q]^T$ is the unit normal vector at $\boldsymbol{\sigma}$ on the loading surface controlling the direction of loading, $\mathbf{m} = [m_p, m_q]^T$ is the unit direction of plastic flow at $\boldsymbol{\sigma}$ and h is the hardening modulus. A superscripted e or p denotes elastic or plastic components of the strains, respectively.

2.3 The critical state

The critical state acts as a reference condition towards which all states approach with increasing shear strain. The critical state therefore has an important role in the model and consideration must be given to the effects of unsaturation in its definition.

It is often assumed that the critical state line (CSL) is straight in the $\upsilon \sim \ln p'$ and $q \sim p'$ planes for both saturated clays and sands. However, there is experimental evidence that the CSL for sands in the $\upsilon \sim \ln p'$ plane may not be straight as it is affected by particle crushing at high stresses (Been et al., 1991; Konrad, 1998). In this study the CSL for saturated sands is assumed to take the form of three linear segments (illustrated in Figure 1) as identified by Russell and Khalili (2002) through a review of several sets of experimental data across a wide stress range. The three linear segments are defined by the six material parameters λ_0, Γ_0, υ_{cr}, λ_{cr}, υ_f and $\lambda_f \cdot \lambda_0$ and Γ_0 are the slope of the initial portion of the CSL and its specific volume at $p' = 1$ kPa, respectively; υ_{cr} is the specific volume at the onset of particle crushing; λ_{cr} is the slope during the particle crushing stage; and υ_f and λ_f are the specific volume at the end of crushing and the slope of the CSL at extremely high stresses, respectively.

Furthermore, the experimental data presented in Wheeler and Sivakumar (1995), Cui and Delage (1996) and Gallipoli et al. (2003) for saturated and unsaturated soils when interpreted in an effective stress framework provides evidence that suction influences the location of CSL in the $\upsilon \sim \ln p'$ plane. Loret and Khalili (2000, 2002) attributed this response to suction hardening, an isotropic hardening phenomenon that controls the size of the yield surface (when using conventional plasticity theory) in addition to plastic volumetric strains, and therefore the location of the CSL.

For saturated conditions a general definition of the CSL in the $\upsilon \sim \ln p'$ plane is adopted in the model formulation for simplicity:

$$\upsilon = f_{cs}(p') \qquad (2a)$$

where f_{cs} is a function unique to a given soil that describes the shape of the CSL. General definition of the CSL in the $\upsilon \sim \ln p'$ for unsaturated conditions thus takes the form:

$$\upsilon = f_{cs}(p', s) \qquad (2b)$$

The CSL in the $q \sim p'$ plane will now be discussed. For saturated clays there is much experimental evidence that the stress ratio ($\eta = q/p'$) approaches a constant value at large shear strains in triaxial tests, and therefore the CSL is linear and passes through the origin. Denoting the slope of the CSL, M_{cs}, and adopting the Mohr-Coulomb failure criterion, it can be shown that at the critical state M_{cs} is simply a function of the critical state friction angle (ϕ'_{cs}) according to:

$$M_{cs} = \eta = \frac{6 \sin \phi'_{cs}}{3 - \sin \phi'_{cs}} \qquad (3)$$

where M_{cs} and ϕ'_{cs} are material constants.

For saturated quartz sands, Colliat-Dangus et al. (1988) and Yamamuro and Lade (1996) showed that the stress ratio also approaches a constant value at large shear strains for stresses ranging from 50 kPa to 15000 kPa. Therefore, it is assumed here that the slope of the CSL in the $q \sim p'$ plane is a material constant for saturated quartz sands, irrespective of the amount of particle crushing that the sand may have experienced. Equation (3) therefore also applies.

Khalili et al. (2004) reviewed triaxial compression test data for several saturated and unsaturated soils from four different laboratories. Application of the effective stress approach was demonstrated and, by adopting an appropriate relationship for χ, ϕ'_{cs} and M_{cs} were found to be material constants for saturated and unsaturated conditions irrespective of suction.

2.4 Elasticity

A simple isotropic elastic rule is adopted and is the same as that used for both sands and clays in other constitutive models. The existence of a purely elastic region is ignored in the general formulation such that all deformation is elastic-plastic. One advantage of this is that the nonvertical response of undrained triaxial test results when plotted in the $q \sim p'$ plane can be modelled.

Incremental elastic volume strain accompanies a change in p' according to a linear relationship between υ and $\ln p'$ such that the elastic bulk modulus K is defined as:

$$K = \frac{\dot{p}'}{\dot{\varepsilon}^e_p} = \frac{\upsilon p'}{\kappa} \qquad (4)$$

where κ is a material constant. For triaxial conditions the elastic shear modulus G is then defined as:

$$G = \frac{\dot{q}}{3\dot{\varepsilon}^e_q} = \frac{3(1-2v)}{2(1+v)} K \qquad (5)$$

where v is Poisson's ratio and is also assumed constant.

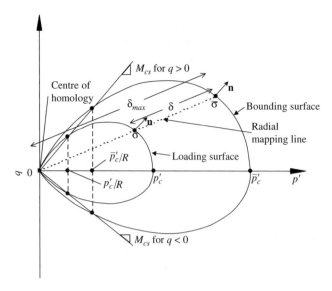

Figure 2. Loading surface, bounding surface, radial mapping line and image point in the $q \sim p'$ plane.

2.5 *Loading surface, bounding surface and image point*

For simplicity, a loading surface and bounding surface of the same shape and homologous about the origin in the $q \sim p'$ plane are assumed. Also, a simple radial mapping line that passes through the origin and σ on the loading surface to intersect the bounding surface at an image point $\bar{\sigma}$ is assumed. It follows that the unit normal vectors of the loading surface and bounding surface at σ and $\bar{\sigma}$ are the same. Simple and versatile functions for the loading surface and bounding surface are adopted and are of the form:

$$f = q - M_{cs} p' \left[\frac{\ln(p'_c/p')}{\ln R} \right]^{1/N} = 0 \qquad F = \bar{q} - M_{cs} \bar{p}' \left[\frac{\ln(\bar{p}'_c/\bar{p}')}{\ln R} \right]^{1/N} = 0 \qquad (6)$$

where the parameters p'_c and $\bar{p}'_c$ control the size of the loading surface and bounding surface, respectively, and represent intercepts with the $q = 0$ axis as illustrated in Figure 2. The material constant R represents the ratio between p' at the intercept of the loading surface with the M_{cs} line and p'_c, and the ratio between $\bar{p}'$ at the intercept of the bounding surface with the M_{cs} line and $\bar{p}'_c$. $\mathbf{n}$ is then defined as:

$$\mathbf{n} = \frac{\frac{\partial f}{\partial \boldsymbol{\sigma}}}{\left\| \frac{\partial f}{\partial \boldsymbol{\sigma}} \right\|} = \left[\frac{-\frac{q}{p'}\left(1 - \frac{1}{N\ln(p'_c/p')}\right)}{\sqrt{\left(-\frac{q}{p'}\left(1 - \frac{1}{N\ln(p'_c/p')}\right)\right)^2 + 1}}, \frac{1}{\sqrt{\left(-\frac{q}{p'}\left(1 - \frac{1}{N\ln(p'_c/p')}\right)\right)^2 + 1}} \right]^T \qquad (7)$$

Note that due to radial symmetry of the loading and bounding surfaces that $\mathbf{n}$ can also be expressed in terms of the stress invariants associated with the bounding surface.

2.6 *Limiting isotropic compression line*

Inherent in the definitions of the bounding surface and elasticity is the existence of a limiting isotropic compression line (LICL), located at a constant shift from the CSL along the κ line in the $\upsilon \sim \ln p'$ plane, towards which the stress trajectories of all isotropic compression load paths approach.

It is for this reason that the CSL and LICL must be correctly defined at high stresses where particle crushing may occur even if high stresses are not of particular interest in the investigation at hand. For saturated conditions the LICL is expressed as:

$$v = f_{cs}\left(\frac{p'}{R}\right) - \kappa \ln R \tag{8}$$

in which R represents the ratio between $\bar{p}'_c$ and p'_c on the κ line passing through the current state.

2.7 Plastic potential

A non-associated flow rule is assumed and is of the same general form used for both sands and clays in other constitutive models. It is ensured that plastic volumetric strains are zero at the stress ratio corresponding to critical state conditions. Consideration is given to the varied plastic deformation mechanisms that dissipate energy. It is expressed as:

$$d = \frac{\dot{\varepsilon}^p_p}{\dot{\varepsilon}^p_q} = \frac{g_{,p'}}{g_{,q}} = M - \eta \tag{9}$$

where d is dilatency and M controls the amount of energy dissipation. It is integrated to give the plastic potential $g = 0$. $\mathbf{m}$ is then defined by the general equation:

$$\mathbf{m} = \frac{\frac{\partial g}{\partial \boldsymbol{\sigma}}}{\left\|\frac{\partial g}{\partial \boldsymbol{\sigma}}\right\|} = \left[\frac{d}{\sqrt{1+d^2}}, \frac{1}{\sqrt{1+d^2}}\right]^T \tag{10}$$

The plastic potential does not intersect the p' axis at right angles and a discontinuity in the slope exists as q changes between slightly positive and slightly negative values. To avoid the development of plastic deviatoric strains during isotropic compression it is assumed that due to a natural rounding process if $q = 0$ then $\mathbf{m} = [1, 0]^T$.

By assuming energy is dissipated purely in friction, with a friction angle equal to that at the critical state, M can be replaced with M_{cs} and the plastic potential of Cam-Clay is recovered. However, other energy dissipation mechanisms such as particle rotation and crushing must also be taken into account and the only restriction on M is that it must be equal to M_{cs} at the critical state.

2.8 Hardening rule

The hardening modulus h is split into two components:

$$h = h_b + h_f \tag{11}$$

where h_b is the modulus at $\bar{\sigma}$ on the bounding surface. h_f is some arbitrary modulus at σ, defined as a function of some distance between $\bar{\sigma}$ and σ, attaining a value of zero at $\bar{\sigma} = \sigma$.

It is assumed $\bar{p}'_c$ undergoes isotropic hardening with changes in suction as well as plastic volumetric strains, similar to the assumptions of Loret and Khalili (2000, 2002). Suction hardening occurs as a result of $\bar{p}'_c$ increasing at a faster rate than $\bar{p}'$ during an increase in suction. It follows that:

$$h_b = -F_{,\bar{p}'_c}\left(\bar{p}'_{c,\varepsilon^p_p} + \bar{p}'_{c,s}\frac{\dot{s}}{\dot{\varepsilon}^p_p}\right)\frac{m_p}{\left\|\frac{\partial F}{\partial \bar{\boldsymbol{\sigma}}}\right\|} \tag{12}$$

Specifically, it is assumed that the LICL undergoes a suction dependant shift $\hat{\gamma}(s)$ along the κ line in the $v \sim p'$ plane. Such a rule describes the situation where suction has an additive effect on the hardening parameter. $\hat{\gamma}(s)$ is positive, has units of stress, and subject to $\hat{\gamma}(s)=0$ when $s \leq s_e$, where s_e is the suction value separating saturated and unsaturated states. In particular, it is referred to as the air entry value (s_{ae}) and air expulsion value (s_{ex}) for drying or wetting paths, respectively. The LICL is then expressed as:

$$\upsilon = f_{cs}\left(\frac{p'-\hat{\gamma}(s)}{R}\right) - \kappa \ln\left(\frac{Rp'}{p'-\hat{\gamma}(s)}\right) \tag{13}$$

It is noted that the shift of the LICL is accompanied by a shift of the CSL, assuming R is a material constant, and the CSL is expressed as:

$$\upsilon = f_{cs}\left(p'-\frac{\hat{\gamma}(s)}{R}\right) - \kappa \ln\left(\frac{p'}{p'-\hat{\gamma}(s)/R}\right) \tag{14}$$

Using the procedure outlined by Loret and Khalili (2000, 2002) it can be shown that the expression linking the plastic volumetric strain that occurs as the hardening parameter moves from the saturated LICL ($\bar{p}'_{c0}$) to the unsaturated LICL ($\bar{p}'_c$) is:

$$\bar{p}'_c = \bar{p}'_{c0}\exp\left[\frac{\upsilon\varepsilon^p_p}{\hat{\lambda}^*-\kappa}\right] + \hat{\gamma}(s) \tag{15}$$

where $\hat{\lambda}^*$ is a function of the slope of the unsaturated LICL at $\bar{p}'_c$. The rate form for $\bar{p}'_c$ then becomes:

$$\dot{\bar{p}}'_c = \left(\frac{\upsilon\bar{p}'_{c0}}{\hat{\lambda}^*-\kappa}\exp\left[\frac{\upsilon\varepsilon^p_p}{\hat{\lambda}^*-\kappa}\right]\right)\dot{\varepsilon}^p_p + \left(\frac{\partial\hat{\gamma}(s)}{\partial s}\right)\dot{s} \tag{16}$$

Other hardening rules may be adopted such that of Loret and Khalili (2000) where suction has a multiplicative effect or combined additive and multiplicative effect on the hardening parameter.

As indicated previously, h_f may take any arbitrary form provided it is equal to zero at $\sigma=\bar{\sigma}$. In the present study, h_f is assumed to be of the form:

$$h_f = k_m \frac{(\delta)}{(\delta_{max}-\delta)}\frac{\Pi}{\bar{p}'_c}p' \tag{17}$$

where

$$\Pi = \bar{p}'_{c,\varepsilon^p_p} + \bar{p}'_{c,s}\frac{\dot{s}}{\dot{\varepsilon}^p_p} \tag{18}$$

and δ_{max} and δ are the distances from the centre of homology and current stress state to the image point respectively as shown in Figure 2. Notice that if $(\delta_{max}-\delta)\to 0$ then $h_f=\infty$ and if $\delta=0$ then $h_f=0$. Also, $k_m \geq 0$ is a scaling parameter controlling the steepness of the response in the $q\sim\varepsilon_q$ plane and is unique for a given soil and may depend on initial conditions. The ratio $\delta/(\delta_{max}-\delta)$ may be written in the alternate form $(\bar{p}'_c - p'_c)/p'_c$ due to radial symmetry of the loading and bounding surfaces. The ratio $\Pi/\bar{p}'_c$ introduces the similarity between h_f and h_b while p' provides a further scaling effect and gives h_f the dimension of stress. It can be seen that h_f is positive until it becomes zero at $\bar{p}'_c = p'_c$. Also, h_b is positive for $\eta/M < 1$, negative for $\eta/M > 1$, and zero when $\eta = M$.

It is possible for h to become zero at some point when $\eta/M > 1$ where h_f and h_b are equal in magnitude and have opposite sign. At this point transition from hardening to softening occurs. Such a situation is observed in the drained stress–strain behaviour of dense sands or heavily over-consolidated clays. However, if h_b and h remain positive, hardening will occur at all times. Such a situation is observed in the drained stress–strain behaviour of loose sands and normally or lightly over-consolidated clays. Additionally, h will become zero when the two requirements $\eta = M = M_{cs}$ and $p'_c = \bar{p}'_c$ are satisfied simultaneously. This situation occurs at the critical state.

Suction hardening rules of the type adopted here enable the prediction of volumetric collapse upon wetting as detailed by Loret and Khalili (2000, 2002). In general terms, wetting causes a reduction in suction and the hardening parameter is forced to retreat to the LICL corresponding to the lower value of suction. Under a limiting confining pressure, this retreat is accompanied by a reduction in volume.

2.9 Coupling the solid, air and water phases

If s is constant during soil deformation, the volume of water in the sample will vary, as will the volume of air. Conversely, s will vary during deformation if either the air or water volumes are held constant. It is therefore necessary to couple s with these volumes. The procedure for coupling the three phases of an unsaturated material in an effective stress framework is detailed by Khalili et al. (2000) and the basic features are repeated here. Note that $\upsilon_w = S_r e$ is the specific pore water volume and $\upsilon_a = (1 - S_r)e$ is the specific pore air volume.

By adopting Betti's reciprocal rule it can be shown that for a material with incompressible grains:

$$-\frac{\dot{\upsilon}_w}{\upsilon} = -\psi\frac{\dot{\upsilon}}{\upsilon} + [c'_m - \psi^2 c]\dot{s} \qquad -\frac{\dot{\upsilon}_a}{\upsilon} = (\psi - 1)\frac{\dot{\upsilon}}{\upsilon} + [\psi^2 c - c'_m]\dot{s} \qquad (19)$$

where $c = -(\partial \upsilon/\upsilon)/\partial p_n$ is the drained compressibility of the soil skeleton ($1/K$) when $\dot{s} = 0$, $c_m = -(\partial \upsilon/\upsilon)/\partial s$ is the compressibility of the soil skeleton with respect to s when $\dot{p}_n = 0$, c'_m is the compressibility of the water phase with respect to s and is dependant on the assumed soil water characteristic curve (SWCC), and $\psi = \partial(\chi s)/\partial s$.

3 MODEL CALIBRATION

3.1 Speswhite kaolin

The model was calibrated for speswhite kaolin in saturated and unsaturated states by Russell (2004) using a similar procedure to that of Loret and Khalili (2002) and the results of isotropic compression and triaxial compression tests reported in Wheeler and Sivakumar (1995).

Russell (2004) showed that the saturated CSL closely fits a single linear segment in the $\upsilon \sim \ln p'$ plane defined by $\lambda_0 = 0.125$ and $\Gamma_0 = 2.588$ (Figure 1). Also, $\phi'_{cs} = 21.9^0$ was suitably described using the χ relationship of Khalili and Khabbaz (1998):

$$\chi = \left[\frac{s}{s_e}\right]^{-0.55} \qquad (20)$$

A value of $s_e = s_{ae} = 85$ kPa was found to fit the data well. The elastic parameters were found to be $\kappa = 0.015$ and $\nu = 0.45$ and values of $N = 1.4$ and $R = 1.6$ was found to be appropriate. A suitable function for M was of the form:

$$M = \left[1.7\left\langle\frac{Rp'}{p'_c} - 1\right\rangle + 1\right]M_{cs} \qquad (21)$$

and satisfies the condition $M = M_{cs}$ at the critical state when $Rp' = p'_c = \bar{p}'_c$. The symbol $\langle \rangle$ is used such that $\langle x \rangle = x$ when $x > 0$ and $\langle x \rangle = 0$ when $x \leq 0$. The k_m parameter of the hardening modulus was defined as:

$$\log_{10} k_m = \left[1.45\left(\frac{p'_0 R}{\bar{p}'_{c0}}\right) - 0.32\right] \qquad (22)$$

Note that these parameters were obtained from results of tests conducted on lightly over-consolidated material. However, they were carefully selected such that stress–strain behaviour typical of heavily over-consolidated material can be simulated.

A suction dependant shift of the CSLs and LICLs in the $\upsilon \sim \ln p'$ plane was observed and defined by:

$$\hat{\gamma}(s) = \begin{cases} \left[45.11\ln\left(\dfrac{s}{34}\right)\right]^{1.25} & \text{for} \quad 100 \leq s \\ 128.5\left(\dfrac{s-s_e}{100-s_e}\right) & \text{for} \quad s_e \leq s < 100 \end{cases} \quad (23)$$

indicating that presence of suction hardening.

3.2 Kurnell sand

The model was calibrated for saturated Kurnell sand, a predominantly quartz sand, by Russell (2004) using a series of drained and undrained triaxial compression tests, as well as isotropic and oedometric compression tests, conducted at stress levels ranging from 10 kPa to 15000 kPa being sufficient for particle crushing to occur. The model was also calibrated for unsaturated Kurnell sand in that investigation using triaxial compression tests. Constant s and constant υ_w conditions were imposed. Also, oedometric compression tests were performed in which suction was held constant. The SWCCs for a range of void ratios were also determined using the filter paper method and pressure plate method. Specific details of experimental procedures and sample preparation are given in Russell (2004). Enlarged lubricated ends were used in all triaxial tests.

Russell and Khalili (2004a, 2004b) showed that the saturated CSL closely fits the three linear segments in the $\upsilon \sim \ln p'$ plane defined by $\lambda_0 = 0.0284$, $\Gamma_0 = 2.0373$, $\upsilon_{cr} = 1.835$, $\lambda_{cr} = 0.195$, $\upsilon_f = 1.25$ and $\lambda_f = 0.04$ (Figure 1). Specific details of f_{cs} are given in those papers. Also, $\phi'_{cs} = 36.3^0$ was found to fit the data well. The elastic parameters were found to be $\kappa = 0.006$ and $\nu = 0.3$ and values of $N = 3$ and $R = 7.3$ was found to be appropriate. A suitable expression for M was found to be:

$$M = [1 + k_d \xi] M_{cs} \quad (24)$$

with k_d defined as:

$$k_d = \dfrac{[-9.6\zeta_0 - 4.1]\langle -\xi_0 \rangle + [23\zeta_0 + 3.4]\langle \xi_0 \rangle}{\xi_0} \quad (25)$$

A suitable expression for k_m, controlling the magnitude of h_f, was found to be:

$$k_m = \dfrac{\langle 52 - 48\exp(\xi_0)\rangle (p'_0)^{0.44}}{1000} \quad (26)$$

The subscript 0 indicates the initial condition of the subscripted variable. The soil water characteristic curves indicated an air entry value of $s_{ae} = 6$ kPa for all void ratios tested, ranging from 0.68 to 0.78. A slight modification of the Khalili and Khabbaz (1998) χ relationship was adopted:

$$\chi = \begin{cases} \left(\dfrac{s}{s_e}\right)^{-0.55} & \text{for} \quad \dfrac{s}{s_e} \leq 25 \\ 25^{0.45}\left(\dfrac{s}{s_e}\right)^{-1} & \text{for} \quad \dfrac{s}{s_e} > 25 \end{cases} \quad (27)$$

The trajectories of the unsaturated triaxial tests approached the saturated critical state line in the $v \sim \ln p'$ plane suggesting it to be unique for the saturated and unsaturated conditions. The experimental results also showed no suction hardening for the soil tested, i.e. $\partial \bar{p}'_c / \partial s = 0$ and $\hat{\gamma}(s) = 0$.

4 MODEL SIMULATIONS OF TEST RESULTS

4.1 Speswhite kaolin

Model simulations of some the triaxial tests results are shown here in Figures 3–5. Note the much improved fit between simulation and experiment compared to that of other constitutive models developed using conventional Cam-Clay based plasticity theory and calibrated using the same set of data (for example Wheeler and Sivakumar, 1995; Loret and Khalili, 2002). The improved fit highlights the versatility of bounding surface plasticity theory. The improvements are attributed to definition of a more realistic loading direction (controlled by the loading surface) and plastic potential than possible when the fundamental principles of conventional Cam-Clay based plasticity theory were followed, including less freedom in defining the direction of loading and associated flow.

4.2 Kurnell sand

Model simulations of some the triaxial test results and oedometric compression test results are shown here in Figures 6–8 in $q \sim \varepsilon_q$ and $\varepsilon_p \sim \varepsilon_q$ planes. Classical stress–strain behaviour is observed. Specifically, hardening occurs up to a peak in the shear resistance, accompanied by initial volumetric

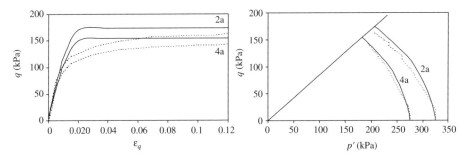

Figure 3. Experimental results and model simulations for triaxial compression tests on speswhite kaolin (after Wheeler and Sivakumar, 1995) in which v and s were held constant, presented in the $q \sim \varepsilon_q$ and $q \sim p'$ planes, with the initial conditions $p_{n0} = 200$ kPa, $s_0 = 200$ kPa and $v_0 = 2.067$ (2a); and $p_{n0} = 150$ kPa, $s_0 = 200$ kPa and $v_0 = 2.127$ (4a).

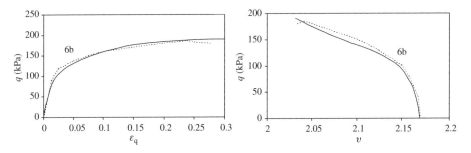

Figure 4. Experimental results and model simulations for a triaxial compression test on speswhite kaolin (after Wheeler and Sivakumar, 1995) in which p_n and s were held constant, presented in the $q \sim \varepsilon_q$ and $q \sim v$ planes, with the initial conditions $p_{n0} = 100$ kPa, $s_0 = 200$ kPa and $v_0 = 2.170$ (6b).

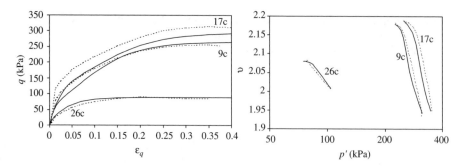

Figure 5. Experimental results and model simulations for triaxial compression tests on speswhite kaolin (after Wheeler and Sivakumar, 1995) in which $\partial q/\partial p_n = 3$ and s was held constant, presented in the $q \sim \varepsilon_q$ and $\upsilon \sim \ln p'$ planes, with the initial conditions $p_{n0} = 100$ kPa, $s_0 = 200$ kPa and $\upsilon_0 = 2.180$ (9c); $p_{n0} = 100$ kPa, $s_0 = 300$ kPa and $\upsilon_0 = 2.188$ (17c) and $p_{n0} = 75$ kPa, $s_0 = 0$ kPa and $\upsilon_0 = 2.080$ (26c).

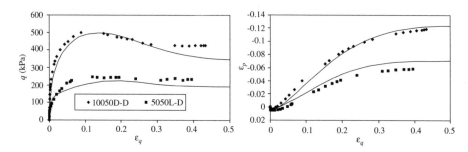

Figure 6. Experimental results and model simulations for triaxial compression tests on Kurnell sand (after Russell, 2004) in which $\partial q/\partial p_n = 3$ and s was held constant, presented in the $q \sim \varepsilon_q$ and $\varepsilon_p \sim \varepsilon_q$ planes, with the initial conditions $p_{n0} = 50$ kPa, $s_0 = 51$ kPa and $\upsilon_0 = 1.770$ (5050L-D); and $p_{n0} = 102$ kPa, $s_0 = 51$ kPa and $\upsilon_0 = 1.658$ (10050L-D).

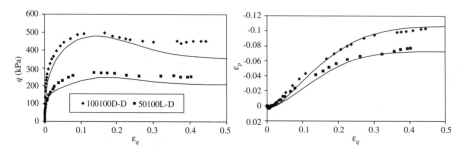

Figure 7. Experimental results and model simulations for triaxial compression tests on Kurnell sand (after Russell, 2004) in which $\partial q/\partial p_n = 3$ and s was held constant, presented in the $q \sim \varepsilon_q$ and $\varepsilon_p \sim \varepsilon_q$ planes, with the initial conditions $p_{n0} = 50$ kPa, $s_0 = 100$ kPa and $\upsilon_0 = 1.763$ (50100L-D); and $p_{n0} = 100$ kPa, $s_0 = 100$ kPa and $\upsilon_0 = 1.687$ (100100L-D).

contraction followed by volumetric expansion. Softening towards the critical state line is observed after reaching the peak and is accompanied by volumetric expansion. Figure 9 shows the oedometric compression test results plotted in $\upsilon \sim \ln \sigma_{1n}$ and $\upsilon \sim \ln \sigma_1'$ planes. Note that it was necessary to assume initial values to make theoretical predictions as the critical state line is defined in a semi-logarithmic plane where zero stress level is undefined. Specifically, initial values of $p_0' = 10$ kPa and $\sigma_3'/\sigma_1' = v/(1 - v) = 0.429$ were assumed, as were elastic strains from $p' = 0$ to p_0'.

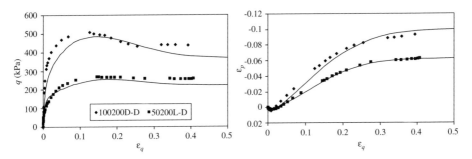

Figure 8. Experimental results and model simulations for triaxial compression tests on Kurnell sand (after Russell, 2004) in which $\partial q/\partial p_n = 3$ and s was held constant, presented in the $q\sim\varepsilon_q$ and $\varepsilon_p\sim\varepsilon_q$ planes, with the initial conditions $p_{n0} = 51$ kPa, $s_0 = 198$ kPa and $\upsilon_0 = 1.780$ (50200L-D); and $p_{n0} = 101$ kPa, $s_0 = 200$ kPa and $\upsilon_0 = 1.697$ (100200L-D).

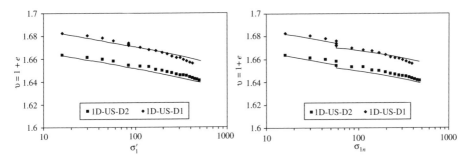

Figure 9. Experimental results and model simulations for oedometric compression tests on Kurnell sand (after Russell, 2004) in which s was held constant, presented in the $\upsilon\sim\sigma_1'$ and $\upsilon\sim\sigma_{1n}$ planes, with the conditions $\upsilon_0 = 1.682$, $p_0' = 10$ kPa and s increased from 0 to 600 KPa at $\sigma_{1n} = 57.4$ KPa (1D-US-D1); and $\upsilon_0 = 1.663$, $p_0' = 10$ kPa and s increased from 0 to 200 kPa at $\sigma_{1n} = 57.4$ kPa (1D-US-D2).

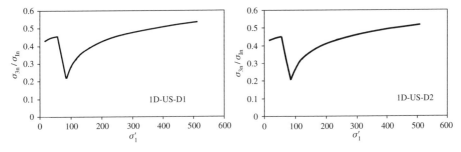

Figure 10. Model simulations for oedometric compression tests on Kurnell sand (after Russell, 2004) in which s was held constant, presented in the $\sigma_{3n}/\sigma_{1n}\sim\sigma_1'$ plane, with the conditions $\upsilon_0 = 1.682$, $p_0' = 10$ kPa and s increased from 0 to 600 kPa at $\sigma_{1n} = 57.4$ kPa (1D-US-D1); and $\upsilon_0 = 1.663$, $p_0' = 10$ kPa and s increased from 0 to 200 kPa at $\sigma_{1n} = 57.4$ kPa (1D-US-D2).

An important point regarding oedometric loading of unsaturated soils was made by Khalili *et al.* (2004). It is possible that suction, which acts isotropically within the soil, may be sufficiently large such that the net radial stress becomes zero, meaning that the soil separates from the confining ring and the condition of zero radial strain does not apply. This was checked through theoretical predictions and, as can be seen in Figure 10, σ_{3n} is always larger than 0 such that the zero radial strain of oedometric compression always applies.

5 CONCLUSIONS

A bounding surface plasticity model has been presented in a critical state framework using the concepts of effective stress. Appealing features of the model include its suitability to describe the stress–strain behaviour of all types of unsaturated soils subjected to various load paths using separate sets of suction independent material parameters.

REFERENCES

Been, K., Jefferies, M.G. & Hachey, J. 1991. The critical state of sands. *Geotechnique*, 41(3): 365–381.
Bishop, A.W. 1959. The principle of effective stress. *Teknisk Ukeblad*, 106(39): 859–863.
Colliat-Dangus, J.L., Desrues, J. & Foray, P. 1988. Triaxial testing of granular soil under elevated cell pressure. *Advanced Triaxial Testing of Soil and Rock*, (R.T. Donage, R.C. Chaney & M.L. Silver ed.) ASTM. 290–310.
Cui, Y.J. & Delage, P. 1996. Yielding and plastic behaviour of an unsaturated compacted silt. *Geotechnique*, 46(2): 291–311.
Dafalias, Y.F. 1986 Bounding surface plasticity. I: Mathematical foundation and hypoplasticity. *Journal of Engineering Mechanics, ASCE*, 112(9): 966–987.
Dafalias, Y.F. & Popov, E.P. 1975. A model for nonlinearly hardening materials for complex loading. *Acta Mechanica*, 21: 173–192.
Gallipoli, D., Gens, A., Sharma, R. & Vaunat, J. 2003. An elasto-plastic model for unsaturated soil incorporating the effects of suction and degree of saturation on mechanical behaviour. *Geotechnique*, 53(1): 123–135.
Khalili, N. & Khabbaz, M.H. 1998. A unique relationship for χ for the determination of the shear strength of unsaturated soils. *Geotechnique*, 48: 681–687.
Khalili, N., Geiser, F. & Blight, G.E. 2004. Effective stress in unsaturated soils, a review with new evidence. *International Journal of Geomechanics*, (*in press*).
Khalili, N., Khabbaz, M.H. & Valliappan, S. 2000. An effective stress based numerical model for hydro-mechanical analysis in unsaturated porous media. *Computational mechanics*, 26: 174–184.
Konrad, J.M. 1998. Sand state from cone penetrometer tests: a framework considering grain crushing stress. *Geotechnique*, 48(2): 201–215.
Loret, B. & Khalili, N. 2000. A three-phase model for unsaturated soils. *International Journal for Numerical and Analytical Methods in Geomechanics*, 24: 893–927.
Loret, B. & Khalili, N. 2002. An effective stress elastic-plastic model for unsaturated porous media. *Mechanics of Materials*, 34: 97–116.
Russell, A.R. & Khalili, N. 2002. Drained cavity expansion in sands exhibiting particle crushing. *International Journal for Numerical and Analytical Methods in Geomechanics*, 26: 323–340.
Russell, A.R. & Khalili, N. 2004a. A bounding surface plasticity model for sands in an unsaturated state. In *Proceedings of the International Conference: From Experimental Evidence towards Numerical Modelling of Unsaturated Soils*, (*in press*).
Russell, A.R. & Khalili, N. 2004b. A bounding surface plasticity model for sands exhibiting particle crushing. *Canadian Geotechnical Journal*, (*in press*).
Russell, A.R. 2004. Cavity expansion in unsaturated soils. PhD thesis, The University of New South Wales, Australia, (*submitted*).
Wheeler, S.J. & Sivakumar, V. 1995. An elasto-plastic critical state framework for unsaturated soil. *Geotechnique*, 45(1): 35–53.
Yamamuro, J.A. & Lade, P.V. 1996. Drained sand behavior in axisymmetric tests at high pressures. *Journal of Geotechnical Engineering, ASCE*, 122(2): 109–119.

Modelling the *THM* behaviour of unsaturated expansive soils using a double-structure formulation

M. Sánchez, A. Gens & S. Olivella
Geotechnical Engineering Department, Technical University of Catalonia, Barcelona, Spain

ABSTRACT: The study of engineering problems in porous media is generally dealt with assuming that they possess a continuous distribution of one type of voids. However, there are some media in which, for a proper handling of the problem, it is crucial to consider the different structural levels involved in the material fabric. This work presents a mathematical formulation that considers the mechanical, hydraulic and thermal problems in a fully coupled way. The Thermo-Hydro-Mechanical (*THM*) approach has been developed to handle engineering problems in porous media with two dominant structures of voids. The formulation allows the consideration of non-isothermal multiphase flows in both media, coupled with the mechanical and the thermal problems. The double porosity approach has been implemented in a finite element code and it has been used to analyze a variety of engineering problems. Special attention has been placed on the analysis and design of high level radioactive waste disposals.

1 INTRODUCTION

The use of expansive clay as a buffer in the design of high level radioactive waste disposals is perhaps the main motivation of many investigations tending to explore its *THM* behavior. In the last few years, more specific tests have been performed leading to a better understanding of swelling clays behavior. Particularly helpful have been the works in which information related to the clay fabric has been provided, revealing a strong influence of the pore structure on the *THM* behavior of expansive materials (i.e. Villar 2000, Cui et al. 2001, Lloret et al. 2003).

Focusing on the problem of radioactive waste repository, most conceptual designs envisage placing the canisters, containing the nuclear waste, in horizontal drifts or vertical boreholes in deep geological media. The empty space surrounding the canisters is filled by an engineered barrier often made up of compacted swelling clays. In this multi-barrier disposal concept both, the geological barrier (host rock) and the engineering barrier (backfill) should be media with very low permeability in order to achieve the required degree of waste isolation. Other functions that the barriers would accomplish are: to provide mechanical stability for the waste canister (by absorbing stresses and deformations) and to seal discontinuities in the emplacement boreholes and drifts, among others.

Significant *THM* phenomena take place in the engineering barrier and in the near field due to the combined actions of the heating arising from the canister, and the hydration coming from the surrounding rock. A crucial aspect of this problem is that many of the *THM* phenomena occurring simultaneously, are strongly *coupled* and interact with each other in a complex way. Therefore, coupled analyses of the relevant *THM* phenomena are generally required to achieve a good understanding of these problems. Coupled *THM* formulations, and the numerical codes built on them, have been widely used in the design and performance assessment studies of nuclear waste disposal (e.g. Olivella et al. 1994, Thomas & He 1995, Gens et al. 1998).

A common feature of the approaches cited above is the assumption of the porous medium as a single porosity material and the adoption of average properties over the elementary representative volume. However, in many cases, the low permeability media is characterized by the presence of more than one kind of voids. For instance, in many compacted soils the fabric is composed by an

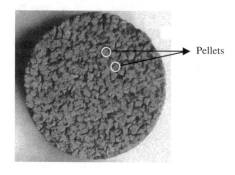

Figure 1. (a) Picture of a sample made up from a mixture of clay pellets (Alonso & Alcoverro 1999). (b) Micrograph of a compacted bentonite sample at a dry density of 1.72 Mg/m^3 (Lloret et al. 2003).

assembly of quasi-saturated aggregates forming a rather open structure that must be distinguished from the clay microstructure itself (Figure 1a). In other cases, the double structure is directly related to the material used, for example in seals composed of high-density pellets with or without powder that fills the pore spaces between them (Figure 1b). Additionally, the necessity to improve single porosity models can also be found in other fields, such as: in the study of the consolidation in fissured clays, in the exploitation of freshwater-bearing reservoirs, in geothermal system and in the study of petroleum reservoirs in stratified or fractured media. In all of these cases the inclusion of the different voids levels in the analyses plays a crucial role to a better understanding and explanation of the material behavior and, evidently, to a proper modeling of these problems. Therefore, two main objectives have been proposed: the first one is the development of a mathematical approach for double porosity media and, the second one, is its experimental validation.

In this work, a quite simple coupled *THM* framework for double porosity media is introduced. The formulation is general, but it has been mainly oriented to the analysis of expansive materials. The mechanical constitutive law is a key element in the modeling of swelling clay behavior and it can be viewed as the nucleus of the double structure formulation for these materials. The developed mechanical model is built on a conceptual framework for expansive soils in which the fundamental characteristic is the explicit consideration of two pore levels (Gens & Alonso 1992). The double structure approach has been implemented in a finite element code and it has been used to explain and reproduce a variety of engineering problems (Sánchez 2004).

2 *THM* FORMULATION

A macroscopic approach developed in the context of the continuum theory for porous media is presented herein. It is assumed that the porous medium is made up of three phases: solid, liquid and gas. The liquid phase contains water and dissolved air whereas the gas phase is made up of dry air and water vapor (the dry air is considered as a single species in spite of the fact that it is a mixture of gasses). The formulation incorporates basic thermal, hydraulic and mechanical phenomena, which are briefly summarized as follows:

- Thermal: heat conduction, heat advection for all phases, and phase changes.
- Hydraulic: liquid advection, gas advection, water vapor and dissolved air diffusion, water evaporation and air dissolution.
- Mechanical: dependence of strains on stresses, suction and temperature changes.

The problem is approached using a multi-phase, multi-species formulation that expresses mathematically the main *THM* phenomena cited above in terms of:

- Balance equations.
- Constitutive equations.
- Equilibrium restrictions.

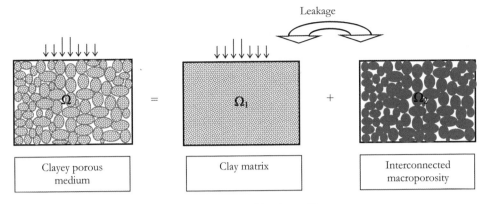

Figure 2. Schematic representation of the two structural levels considers.

Concepts of double porosity theory (i.e. Barenblatt et al. 1960, Aifantis 1980, Huayakorn et al. 1983, Ghafouri & Lewis 1996) has been used to extend an existing *THM* formulation for single porosity media (Olivella et al. 1994) to media with two structures of pores. Double porosity theory considers the porous medium as two interacting continuous media coupled through a leakage term. This term controls the mass transfer between the two porous media. A good schematic representation of the double porosity theory is presented in Ghafouri & Lewis (1996). Figure 2 presents a similar conceptual model that could be adopted for some expansive materials with two different structures of pores. For instance, in the mixture of clays pellets showed in Figure 1b, the medium 1 could be related to the solid particles of the clay pellets and the voids inside the pellets, and the medium 2 could be associated with the macrostructure formed by the pellets (as a whole) and the macropores between pellets. Porosity, fluids pressures, permeability, degree of saturation and other properties are considered separately for each continuum. It is assumed that two structures of interconnected pores exist, with different properties and fluids that flow through them.

The main aspects of the formulation are summarized as follows:

- Two distinct porous media have been considered, with the definitions of different properties in each medium.
- Multiphase, non-isothermal flows in each domain are considered.
- Mass transfer processes between media are controlled through mass transfer terms.
- Stress-small strain constitutive laws can be defined for each porous medium.
- Thermal equilibrium between the phases and the media is assumed.
- The relevant *THM* phenomena are considered in a fully coupled way.

2.1 Balance equations

The compositional approach has been adopted to establish the mass balance equations (Olivella et al. 1994). This approach consists of balancing the species (mineral, water and air) rather than the phases (solid, liquid and gas). In this way the phase change terms do not appear explicitly, which is particularly useful when equilibrium is assumed.

2.1.1 Mass balance equations

The following mass balance equations for double porosity media are considered in the approach (Sánchez 2004):

- Water mass balance equation.
- Air mass balance equation.
- Solid mass balance equation.

2.1.2 *Energy balance equation*

The balance of energy has been expressed in terms of internal energy (Olivella et al. 1994). The most important processes of energy transfer in a porous medium have been considered, which are: conduction, advection and phase change (Sánchez 2004).

In this approach thermal equilibrium between the phases and the media has been assumed, therefore the temperature is the same for the phases and only one equation of total energy is required. This assumption is generally valid in low permeability media. If either the characteristic of the problem or the experimental evidence justifies the necessity of a more detailed treatment of this equation, different temperatures in the two media can be considered.

2.1.3 *Momentum balance (equilibrium)*

The balance of momentum for the porous medium reduces to the equilibrium equation for total stresses, if inertial terms are neglected:

$$\nabla \cdot \sigma_t + \mathbf{b} = 0 \qquad (1)$$

where σ is the stress tensor and **b** the vector of body forces. Through the constitutive model, the equilibrium equation is transformed into a form in terms of the solid velocities, fluid pressures and temperatures (Sánchez 2004). In Section 3, a mechanical constitutive model for expansive clays is introduced.

2.2 *Constitutive equations*

The constitutive equations establish the link between the unknowns and the dependent variables. The governing equations are finally written in terms of the unknowns when the constitutive equations are substituted in the balance equations. Depending on the problem analyzed, specific constitutive laws for the thermal, hydraulic and mechanical problem can be adopted. For instance, in FEBEX Report (2000) and Sánchez (2004) it can be found a description of the main constitutive laws used in the different analyses performed with this formulation.

2.3 *Equilibrium restrictions*

According to Olivella et al. (1994), it is assumed that the phase changes are rapid in relation to the characteristic times typical of these types of problems. Therefore, they can be considered in local equilibrium, giving rise to a set of equilibrium restrictions that must be satisfied at all times. Also, the adopted compositional approach has the advantage that the phase change terms do not appear explicitly and the number of equations is thereby reduced. Equilibrium restrictions are given by the concentration of water vapor in gas phase, which is computed through the psychometric law; and by the concentration of dissolved air in liquid phase, which is evaluated by means of Henry's law.

2.4 *Mass transfer between media*

A simple model for the term related to the mass transfer between media can be expressed as:

$$\Gamma^i = \gamma(\Psi_1 - \Psi_2) \qquad (2)$$

where γ is the leakage parameter and Ψ_j ($j = 1, 2$) represents the thermodynamic force involved in the mass transfer. When the water mass transfer is considered, the total water potential is the variable involved in Equation (2). In some cases, due to the characteristics of the problems, variables related to the total water potential can be adopted as main responsible of the mass transfer process (for instance: suction, water pressure or temperature). Here, this term is presented in a generic form. In each application and according with the specific characteristics of the problem, the thermodynamic force involved in the mass transfer is specified. In Equation (2), it is assumed that the process of mass transfer has reached a quasi-steady state, hence the name of quasi-steady models (Huayakorn et al. 1983). A more refined treatment of this term can be made through the unsteady models (Huayakorn et al. 1983). In these models the transfer of mass between media is obtained solving the 1-D diffusion problem for an idealized geometry of the matrix pores (parallel fracture and prismatic or spherical blocks). In this formulation the two kinds of models can be potentially used, the selection will depend on the characteristic of each problem (Sánchez 2004).

3 MECHANICAL CONSTITUTIVE MODEL

Comprehensive modeling of unsaturated expansive clays is a complex problem. The swelling behavior of these clays has often been reproduced through relatively simple and empirical laws, which relate the material response to suction changes and applied stresses. However, there are relatively a few formulations that integrate the main aspects of behavior in a unified framework (Gens & Alonso 1992, Alonso et al. 1999, Cui et al. 2002). In Gens & Alonso (1992), particular attention is placed on the clay structure and how it can be integrated in the constitutive modeling of expansive soils. The fabric of expansive clays has been actively studied (Villar 2000, Cui et al. 2001, Lloret et al. 2003) detecting a marked double structure. For instance, Figure 3a shows shows the results of mercury intrusion porosimeter tests of FEBEX bentonite (Fig. 2b), in which a clear bimodal pore distribution can be observed. The dominant values are 10 nm that would correspond to the pores inside clay aggregates. Whereas, a larger pore size, which depends on the compaction dry density, ranges from 10 μm and 40 μm. The boundary between the two pore size families is around 130 nm. These two dominant pores size could be associated with two basic structural levels (Figure 3b):

- The macrostructure, composed by arrangements of clay aggregates, as a whole, with macropores between them (medium 2).
- The microstructure, which corresponds to the active clay minerals and their vicinity (medium 1).

The model introduced in this work is based on the general framework proposed by Gens & Alonso (1992) and considers some of the improvements proposed by Alonso et al. (1999). The model presented herein has been formulated using concepts of elasto-plasticity for strain hardening materials. A series of modifications and developments have been performed in order to enhance the constitutive law and also to formulate the model in a more suitable form for its implementation in a finite element code. One of the aims is to provide a more general mathematical framework in order to achieve a more general interpretation of the phenomena that take place in expansive clays when they are subjected to complex *THM* paths. With this objective, concepts of generalized plasticity theory have been included in the formulation of the model.

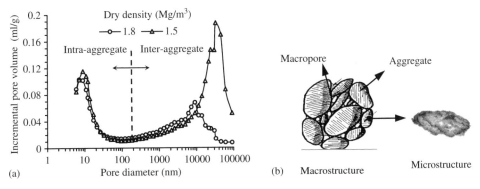

Figure 3. (a) Distribution of incremental pore volume for two compacted bentonite samples at different dry densities (Lloret et al. 2003). (b) Schematic representation of the two structural levels considers.

The mathematical framework of the model is presented in detail in Sánchez (2004) and Sánchez et al. (2004). The Appendix contains the main model equations. The model is formulated in terms of the three stress invariants (p, J, θ), suction (s) and temperature (T).

Finally, the complete model formulation requires the definition of laws for: (i) the macrostructural level, (ii) the microstructural level and (iii) the interaction between the structural levels. These laws are briefly introduced in the following sections.

3.1 Macrostructural model

The inclusion of this structural level in the analysis allows the consideration of phenomena that affect the skeleton of the material, for instance deformations due to loading and collapse. These phenomena have a strong influence on the macroscopic response of expansive materials. The macrostructural behavior can be described by concepts and models of unsaturated nonexpansive soils, such as the elasto-plastic Barcelona Basic Model (*BBM*) (Alonso et al. 1990). The *BBM* considers two independent stress variables to model the unsaturated behavior: the net stress (σ) computed as the excess of the total stresses over the gas pressure ($\sigma_t\text{-}\mathbf{I}p_g$), and the matric suction (s_2), computed as the difference between gas pressure and liquid pressure ($p_{g1} - p_{l2}$). The *BBM* extends the concept of critical state to the unsaturated conditions.

The *BBM* yield surface (Fig. 4a) depends not only on the stress level and on the history variables (as in a critical state model) but also on the matric suction. The *BBM* yield surface (F_{LC}) is given by:

$$F_{LC} = 3J^2 - \left[\frac{g(\theta)}{g(-30°)}\right]^2 M^2 (p + p_s)(p_0 - p) = 0 \quad (3)$$

where M is the slope of the critical state, p_0 is the apparent unsaturated isotropic pre-consolidation pressure, $g(\theta)$ is a function of the lode angle, p is the net mean stress and p_s considers the dependence of shear stress on suction and temperature. A basic point of the model is that the size of the yield surface increases with suction. The trace of the yield function on the isotropic *p-s* plane is called *LC* (Loading-Collapse) yield curve, because it represents the locus of activation of irreversible deformations due to loading increments or collapse (when the suction reduces). The position of the *LC* curve is given by the pre-consolidation yield stress of the saturated state, p_0^* (hardening variable), according to the following expression:

$$\dot{p}_0^* = p_0^* \frac{(1+e)}{(\lambda_{(0)} - \kappa)} \dot{\varepsilon}_v^p \quad (4)$$

where e is the void index, ε_v^p is the volumetric plastic strain, κ is the elastic compression index for changes in p and $\lambda_{(0)}$ is the stiffness parameter for changes in p for virgin states of the soil

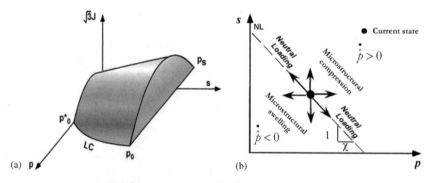

Figure 4. (a) Three dimensional representation of the *BBM* yield surface. (b) Definition of microstructural swelling and contraction directions.

in saturated conditions. The inclusion of the thermal effects has been made according to Gens (1995). In this way it is considered that temperature increases reduce the size of the yield surface and the strength of the material. This is a well-established fact for saturated conditions (Hueckel & Borsetto, 1991) which can also be extended to the unsaturated conditions, as it was confirmed in recent experimental works (Romero et al. 2003, Villar & Lloret 2003).

3.2 Microstructural model

The microstructure is the seat of the basic physical–chemical phenomena occurring at clay particle level. It is assumed that these phenomena are basically reversible (Gens & Alonso 1992). The strains arising from microstructural phenomena are considered elastic and volumetric.

The microstructural effective stress is defined as:

$$\hat{p} = p + \chi s_t; \qquad s_t = s_2 + s_o \qquad (5)$$

where s_0 the osmotic suction. It is assumed that χ is a constant ($\chi > 0$) and that the total suction (s_t) is equal to the matric suction (s_2), because the effect of the osmotic suction is not considered in this work. In Guimarães et al. (2001), the formulation is extended to include geochemical variables. In Equation (5) it is also assumed hydraulic equilibrium between the water potentials of both structural levels (this implies $s = s_1 = s_2$). The extension of the constitutive model to handle problems in which this hypothesis is released is presented in Sanchez (2004).

In the p–s plane the line corresponding to constant microstructural effective stresses is referred to as Neutral Line (NL), since no microstructural deformation occurs when the stress path moves on it (Fig. 4b). The increment of microstructural elastic strains is expressed as:

$$\dot{\varepsilon}^e_{v1} = \dot{\varepsilon}_{v1} = \frac{\dot{\hat{p}}}{K_1} = \frac{\dot{p}}{K_1} + \chi \frac{\dot{s}}{K_1} \qquad (6)$$

where the subscript 1 refers to the microstructural level, the superscript e refers to the elastic component of the volumetric (subscript v) strains and K_1 is the microstructural bulk modulus.

According to (6) the Neutral Line divides the p–s plane into two parts (Fig. 4b), defining two main generalized stress paths, which are identified as:

$\dot{\hat{p}} > 0 \Rightarrow$ microstructural contraction (MC); $\qquad \dot{\hat{p}} < 0 \Rightarrow$ microstructural swelling (MS)

3.3 Interaction between structural levels

In expansive soils there are other mechanisms in addition to the ones included in the BBM which induce plastic strains. This irreversible behavior is ascribed to the interaction between the macro and micro structures (Gens & Alonso 1992). Analyzing the behavior of expansive clays under cycles of suction reversals (e.g. Pousada 1984), two main aspects can be highlighted:

- the irreversible behavior appears independently of the applied suction.
- it is difficult to determine the initiation of the yielding.

These facts aim the use of the generalized plasticity theory to formulate the model (Sánchez et al. 2004). In a generalized plasticity model the yield function is not defined or it is not defined in an explicit way (e.g. Pastor et al. 1990, Lubliner et al. 1993). There are significant advantages in using the generalized plasticity theory to model the plastic mechanism related to the interaction between both pores structures. Some of them are:

- No clear evidence exists concerning the shape of the internal yield surfaces corresponding to the interaction mechanisms between the two structural levels. Furthermore, their experimental determination does not appear to be easy either.
- The effect of drying/wetting cycles on the behavior of expansive soils is a matter of great practical importance. Generalized plasticity is especially well adapted to deal with cyclic loading (Pastor et al. 1990).

- It provides sufficient flexibility to incorporate additional microstructural phenomena such as non-equilibrium microstructural suction (Sánchez 2004), or geochemical variables such as osmotic suction and cation exchange (Guimarães et al. 2001).

Moreover, in the case that the yield surfaces related to this mechanism can be experimentally defined, there are no problems to include them in the modeling, since the classical plasticity theory is a particular case of the generalized plasticity (Lubliner et al. 1993).

It is assumed that the microstructural behavior is not affected by the macrostructure but the opposite is not true, i.e. macrostructural behavior can be affected by microstructural deformations, generally in an irreversible way. An assumption of model is that the irreversible deformations of the macrostructure are proportional to the microstructural strains according to interaction functions f (Gens & Alonso 1992).

The plastic macrostructural strains are evaluated by the following expression:

$$\dot{\varepsilon}^p_{v2} = f\,\dot{\varepsilon}_{v1} \qquad (7)$$

Two interaction functions f are defined: f_c for microstructural contraction paths and f_S for microstructural swelling paths. In the case of isotropic load, the interaction function depends on the ratio p/p_0 (p_0 is the net mean yield stress at current suction and temperature). This ratio is a measure of the degree of openness of the macrostructure. When p/p_0 is low it implies a dense packing of the material. It is expected that under this condition the microstructural swelling (*MS* path) affects strongly the global arrangements of clay aggregates. So, the higher values of the f_s function correspond to low values of p/p_0. In this case the microstructure effects induce a more open macrostructure, which implies a macrostructural softening. On the other hand, when the microstructure contracts (*MC* path) the larger (induced) macrostructural plastic strains occur with open macrostructures (values of p/p_0 close to 1). Under this path the clay tends to a more dense state, which implies a hardening of the macrostructure. This coupling between both plastic mechanisms is considered mathematically assuming that:

$$\dot{\varepsilon}^p_{v2} = \dot{\varepsilon}^p_{vLC} + f\,\dot{\varepsilon}_{v1} \qquad (8)$$

where ε^p_{vLC} is the plastic strains induced by the yielding of the macrostructure (*BBM*). In fact the coupling is given by p_0^*, hardening variable of the macrostructure (Fig. 4a), which depends on the total plastic volumetric strain (4). In this way is considered that microstructural effects can affect the global arrangements of aggregates (macrostructure).

Note that the material response will depend strongly on the direction of the microstructural stress path relative to the *NL*, which delimits two regions of different material behavior. A proper modeling of this behavior requires the definition of specific elasto-plastic laws for each domain, in order to describe correctly the material behavior according to the microstructural stress path followed (*MC* or *MS*). Generalized plasticity theory can deal with such conditions, allowing the consideration of two directions of different behavior and the formulation of proper elasto-plastic laws for each region. A complete model description includes the definition of the: (i) loading and unloading direction, (ii) plastic flow direction, and (iii) a plastic modulus.

In summary, the behaviour of the macrostructure is modeled in the context of classical plasticity (*BBM*). This is a proper framework because the yield surface associated to this behavior could be generally inferred by the usual methodology of classic plasticity. The microstructural effects have been modeled using a nonlinear elastic model. The interaction between pores structures has been model using the more general framework of generalized plasticity theory. Finally, the governing small strain–stress equations have been obtained using a general framework for multidissipative materials. Comprehensive description of these parts of the model is presented in detail in Sánchez (2004) and Sánchez et al. (2004).

A critical step in the implementation of a mechanical model in a finite element program is the development of a proper algorithm to update the stresses and the internal variables of the model.

Since the stresses should be integrated many times on the course of a typical non-linear simulation, the selection of the algorithm has been based on the accuracy of the solution and also on its robustness and efficiency. The numerical integration of the model has been performed using a refined Euler scheme with automatic sub-stepping and error control (Sánchez 2004).

4 NUMERICAL IMPLEMENTATION

The formulation has been implemented in the finite element program CODE_BRIGHT, which is a tool designed to handle coupled *THM* problems in geological media. One unknown (state variable) is associated to each of the balance equations presented. The unknowns are obtained by solving the system of PDE's (Partial Differential Equations) numerically in a coupled way. From state variables, dependent variables are calculated using the constitutive equations or the equilibrium restrictions.

The numerical approach can be viewed as divided into two parts: spatial and temporal discretization. Galerkin finite element method is used for the spatial discretization while finite differences are used for the temporal discretization. Finally, since the problems are in general non-linear, the Newton-Raphson method was adopted as iterative scheme.

5 APPLICATION

The proposed approach has been used to analyze several problems involving coupled *THM* processes in geological media (Sánchez 2004). A good performance of the double structure formulation has been observed in the different cases analyzed.

The applications have been mainly focused on the study of the coupled *THM* phenomena in engineered clay barriers and seals. However, the suggested formulation is general and it is not limited to study problems in which expansive materials are involved. Other kind of problems have been studied, for instance, the problem of consolidation in fissured clays or the simulations of petroleum exploitations in stratified media have also been analyzed (Sánchez 2004). As the main application it can be mentioned the analysis of the heating and hydration of an ongoing large scale heating test. In this test the clay barrier has been constructed with blocks of compacted FEBEX bentonite. A more proper analysis of the effects of the clay fabric changes on the test evolution has been approached using the double structure framework. The analysis has been helpful for the understanding and simulation of the large scale experiment. The mechanical constitute model used in this simulation is presented in the following paragraphs.

In the context of the Febex project a wide experimental program has been performed to study the behavior of the FEBEX bentonite (FEBEX Project 2000). The studies have been oriented towards the physic-mechanical characterization of this clay. Oedometers tests, with suction control, in which a combination of loading paths (up to 10 MPa) at constant suction and wetting and drying paths (up to 550 MPa) at constant load were performed. The experimental study also includes swelling tests under constant volume conditions in order to determine the swelling pressure and the stress path followed during wetting. These two kinds of tests provide the opportunity to examine the behavior of the model over a wide range of stress paths. A detailed description of the tests is presented in Lloret et al. (2003). The model is first applied to analyze the tests coded as S1 and S5 (Fig. 5a) performed over a compacted bentonite of 1.7 Mg/m^3, with an initial gravimetric water content near the 14%. All the tests have the same initial and final suction and load, but their trajectories are very different.

Figure 5b shows the variation of void ratio during the initial stage of suction modifications and subsequent loading. The starting points for the loading stages are very different because of the large dependence of volumetric strains on suction applied at low loads. On loading, the stiffness of the bentonite (i.e. the slope of the void ratio vs. vertical stress line plotted in semi-logarithmic scale) reduces slightly as the suction applied during loading increases. However, the most noticeable effect of suction is the shifting of the point at which there is a change in the slope of these lines (indicated by a vertical arrow in the Fig. 5b). This change is interpreted as the crossing of a yield surface and the load at which it takes place can be considered as an apparent pre-consolidation pressure. Large pre-consolidation pressure reductions are apparent at low suction values. According to the conceptual

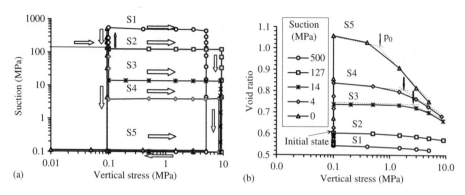

Figure 5. (a) Generalized stress paths followed by tests. (b) Variation of void ratio during the initial stage of suction variation and subsequent loading (Lloret et al. 2003).

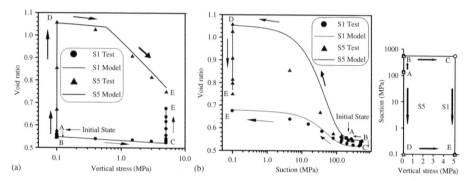

Figure 6. Computed variation of void ratio for S1 and S5 tests. Experimental results are provided for comparison.

model adopted, the reduction of the yield point is due to the irreversible macrostructural strains induced by microstructural deformations that have occurred during the swelling at low suction values. Yielding was not reached in tests S1 and S2, and it is assumed that the pre-consolidation stress, for this density, corresponds approximately to the vertical stress value reached during static compaction, about 18 MPa.

In test S1 the specimen is loaded under a high 550 MPa suction up to a 5.1 MPa vertical load and then is wetted reducing the suction to 0 in stages. In contrast, test S5 is first wetted at a low applied vertical stress value of 0.1 MPa and afterwards the sample, already saturated, is loaded to a vertical stress of 5.0 MPa.

In Figure 6a, it can be observed that major features of behavior are correctly reproduced including:

- Large swelling strains when the material is wetted at low stresses (Path B-D, Test S5), Smaller, but still significant, swelling strains when the soil is wetted under a 5 MPa vertical stress (Path C-E, Test S1).
- Change of the slope of the compression line during loading indicating yield in test S5 (Path D-E). No yield is apparent during the loading at high suction of specimen S1 (Path B-C).
- Final void ratio (point E) is different in the two samples; there is a measure of stress path dependency, at least regarding volumetric strains.

Good reproduction of behavior is also achieved when considering the experimental results in terms of void ratio vs. suction variation (Fig. 6b); although some departures are observed at intermediate stages of the swelling of test S5.

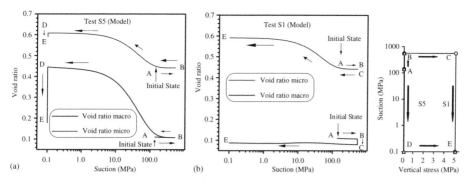

Figure 7. Evolution of the computed macrostructural and microstructural void ratios. Test S5 and S1.

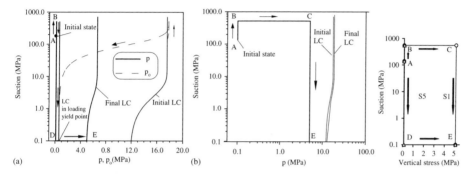

Figure 8. (a) Stress path and successive LC yield surfaces for test S5. (b) Stress path and LC yield surfaces for test S1.

Figure 7a shows the evolution of the microstructural and macrostructural void ratio computed for test S5. During the swelling stage (path B-D), the microstructural strains are relatively large and they cause even larger irreversible strains in the macrostructure because of the large coupling between the two structural levels that exist at low stresses. During the subsequent loading (path D-E) under saturated conditions the deformation of the macrostructure is significant but it is not due to microstructural strains that are now quite small. This part of the test is basically controlled by the behavior of the macrostructure. Some of the positions of the LC yield curve during the performance of test S5 are shown in Figure 8a. It can be seen that during swelling (path B-D) the LC curve moves to the left in response to the irreversible swelling strains taking place in the macrostructure. Subsequent loading (path D-E) takes the LC again to the right to the final load value of 5.0 MPa. Indeed the yield point observed and computed (Fig. 6a) corresponds to the crossing of the LC during this loading stage.

The behavior of sample S1 is quite different (Fig. 7b). During the first stage of drying (path A-B) and subsequent loading (path B-C), the microstructural volumetric strains are very small. The macrostructural strains are also small even during loading because of the high stiffness imparted to the sample by the large 500 MPa suction. During the swelling stage (path C-E) under a 5.1 MPa load, the microstructural strains are significant although smaller than for test S5 because of the higher load applied. The most significant difference is, however, that the macrostructural strains that are induced are quite small because now the stress state is much closer to the LC, i.e. the sample is in a comparatively looser state and the potential for macrostructural disruption is much lower. The loading stage (path B-C) takes place inside the LC yield surface (Fig. 8b), so no yield is expected and none was observed.

Now the basic reason for the stress path dependency of volumetric strains can be readily identified. The basic difference is that in test S5 the large swelling strains take place at low stress values

Table 1. Main parameters of the mechanical constitutive law.

Parameters defining the expansive model

κ	0.005	ζ (MPa^{-1})	1.00	p_c (MPa)	0.50
r	0.90	α_m (MPa^{-1})	$2.1\,e^{-02}$	p_0^* (MPa)	12.0
κ_s	0.001	β_m (MPa^{-1})	$2.3\,e^{-03}$	$\lambda(0)$	0.080

$f_c = 1 + 0.9 \tan h\,(20\,(p/p_0) - 0.25)$ $\qquad f_s = 0.8 - 1.1 \tan h\,(20\,(p/p_0) - 0.25)$

and, consequently, the interaction with the macrostructure is very strong and results in large plastic strains that are not fully recovered upon subsequent loading. In test S1, development of plastic strains in the macrostructure is quite reduced because, when the swelling of the microstructure takes place, the interaction between the two structural levels is small.

It is important to highlight that despite the relatively complexity of the model, only standard tests of non-saturated soil mechanics have been used to identify the great part of the model parameters, which are indicated in Table 1.

6 CONCLUSIONS

An open and general *THM* mathematical approach for porous media with two distinctive types of voids has been proposed. The analyses have been mainly focused on the study of the coupled *THM* process in engineered clay barriers. So, special attention has been placed on the formulation of the constitutive model for expansive soils. In order to be closer to the typical fabric of expansive materials, the existence of two pores structures has been explicitly included in the model. The approach has been implemented in a finite element code program allowing the incorporation of these concepts in the numerical analyses readily. The double structure *THM* approach has revealed a good performance in all of the problems analyzed. In this work the validation of the mechanical constitutive law has been presented.

ACKNOWLEDGEMENTS

The authors are grateful to ENRESA for their support in this work. FEBEX is a research project partially funded by the European Commission. The authors also acknowledge valuable discussions with Antonio Lloret, María Victoria Villar, Xavier Pintado and Eduardo Alonso.

REFERENCES

Aifantis, E. 1980. On the problem of diffusion in solids. *Actha Mechanica* 37(3–4): 265–296.
Alonso, E. & Alcoverro, J. 1999. CATSIUS CLAY Project. Calculation and testing of behaviour of unsaturated clay as barrier in radioactive waste repositories. *Technical public. 11/99*. Madrid: ENRESA.
Alonso, E., Gens, A. & Josa, A. 1990. A constitutive model for partially saturated soils. *Géotechnique*, 40(3): 405–430.
Alonso, E., Vaunat, J. & Gens, A. 1999. Modelling the mechanical behaviour of expansive clays. *Engineering Geology* 54: 173–183.
Barrenbaltt, G., Zeltov, I. & Kochina, N. 1960. Basic concepts in the theory of seepage of homogeneous liquids in fissured rocks. *Pirkl. Mat. Mekh.* 24: 852–864.
Cui, Y., Loiseau, C. & Delage, P. 2001. Water transfer through a confined heavily compacted swelling soil. *6th International Workshop on Key Issues in Waste Isolation Research; Proc. Symp. Paris*, November 200: 43–60.
Cui, Y.J., Yahia-Aissa, M. & Dalage, P. 2002. A model for the volume change behavior of heavily compacted selling clays. *Engineering Geology*, 64: 233–250.
FEBEX Project, 2000. Full-scale engineered barriers experiment for a deep geological repository for high level radioactive waste in crystalline host rock. *Final project report. EUR 19612 EN*, European Commission, Brussels.

Gens, A. 1995. Constitutive Laws. In A. Gens P. Jouanna & B. Schrefler *Modern issues in non-saturated soils*: 129–158. Wien New York: Springer-Verlag.

Gens, A. & Alonso, E.E. 1992. A framework for the behaviour of unsaturated expansive clays. *Can. Geotech. Jnl.*, 29: 1013–1032.

Gens, A., Garcia Molina, A., Olivella, S., Alonso, E.E. & Huertas, F. 1998. Analysis of a full scale in-situ test simulating repository condition. *Int. Jnl. Numer. Anal. Meth. Geomech.*, 22: 515–548.

Ghafouri, H. & Lewis, R. 1996. A finite element double porosity model for heterogeneous deformable porous media. *Int. Jnl. Numer. Anal. Meth. Geomech* 20: 831–844.

Guimaraes L., Gens A., Sánchez M. & Olivella S. 2001. Chemo-mechanical modelling of expansive materials. *6th International Workshop on Key Issues in Waste Isolation Research; Proc. Symp.*: 463–465 November 2001. Paris.

Hueckel, T. & Borsetto, M. 1990. Thermoplasticity of saturated soils and shales: constitutive equations. Journal of Geotechnical Engineering, ASCE, 116(12): 1765–1777.

Huyakorn, B., Lester & Faust., C. 1983. Finite element techniques for modelling groundwater flow in fractured aquifers. *Water Resources Research*, 19(4): 1019–1035.

Lloret, A., Villar, M.V., Sánchez, M., Gens, A., Pintado, X. & Alonso, E. 2003. Mechanical behaviour of heavily compacted bentonite under high suction changes. *Géotechnique*, 53(1): 27–40.

Lubliner, J. & Auricchio, F. 1996. Generalized plasticity and shape-memory alloys. *Int. J Solids Structures*, 33(7): 991–1003.

Olivella, S., Carrera J., Gens, A. & Alonso, E.E. 1994. Non-isothermal multiphase flow of brine and gas through saline media. *Transport in porous media*, 15: 271–293.

Olivella, S., Gens, A., Carrera, J. & Alonso, E.E. 1996. Numerical formulation for a simulator (CODEBRIGHT) for the coupled analysis of saline media. *Engineering Computations*, 13(7): 87–112.

Pastor, M., Zienkiewics, O. & Chan, A. 1990. Generalized plasticity and the modelling of soil behaviour. *Int. Jnl. Numer. Anal. Meth. Geomech.*, 14: 151–190.

Pousada, E. 1984. Deformabilidad de arcillas expansivas bajo succión controlada. *PhD Thesis*, Technical University of Madrid, Spain.

Romero, E., Gens, A. & Lloret, A. 2003. Suction effects on a compacted clay under non-isothermal conditions. *Géotechnique*, 53(1): 65–81.

Sánchez, M. 2004. Thermo-hydro-mechanical coupled analysis in low permeability media. *Ph.D. Thesis*, Technical University of Catalonia. Barcelona.

Sánchez, M., Gens, A., Guimarães, L. & Olivella, S. 2004. A double structure generalized plasticity model for expansive materials. *Int. Jnl. Numer. Anal. Meth. In Geomech.* (accepted).

Thomas, H.R. & He, Y. 1995. An analysis of coupled heat, moisture and air transfer in a deformable unsaturated soil. *Géotechnique*, 45: 667–689.

Villar, M.V. 2000. Thermo-hydro-mechanical characterization of Cabo de Gata Bentonite (in Spanish). *Ph.D. Thesis*, Complutense University, Madrid.

Villar, M.V. & Lloret, A. 2003. Temperature influence on the hydro-mechanical behaviour of a compacted bentonite. *Large scale field tests in granite – Field emplacement and instrumentation techniques, Proc. Symp. Sitges*, 12–14 November 2003.

7 APPENDIX

7.1 MECHANICAL MODEL

A detailed description of the mechanical model can be found in Sánchez (2004) and Sánchez et al. (2004). The *BBM* yield surface (F_{LC}) is given by (3) and the plastic potential (*G*) is expressed as:

$$G = \alpha 3J^2 - \left[\frac{g(\theta)}{g(-30°)}\right]^2 M^2 (p + p_s)(p_0 - p) = 0 \tag{A1}$$

where α is determined according to Alonso et al. (1990).

The dependence of the tensile strength on suction and temperature is given by:

$$p_s = ks \, e^{-\rho \Delta T} \tag{A2}$$

where k and ρ are model parameters.

The dependence of p_0 on suction is given by:

$$p_0 = p_c \left(\frac{p^*_{0T}}{p_c} \right)^{\frac{\lambda_{(0)} - \kappa}{\lambda_{(s)} - \kappa}} ; \qquad p^*_{0T} = p^*_0 + 2\left(\alpha_1 \Delta T + \alpha_3 \Delta T |\Delta T| \right) \tag{A3}$$

where p_c is a reference stress, α_1 and α_3 are models parameters. $\lambda_{(s)}$ is the compressibility parameter for changes in net mean stress for virgin states of the soil. This parameter depends on suction according to:

$$\lambda_{(s)} = \lambda_{(0)} \left[r + (1-r) \exp(-\zeta s) \right] \tag{A4}$$

where r is a parameter which defines the minimum soil compressibility (at infinity suction) and ζ is a parameter which controls the rate of decrease of soil compressibility with suction.

The stress invariants are evaluated as:

$$p = \frac{1}{3}(\sigma_x + \sigma_y + \sigma_z); \quad J^2 = 0.5 \ \text{tr}(s^2); \quad \theta = -\frac{1}{3}\sin^{-1}\left(1.5\sqrt{3} \det s / J^3\right) \tag{A5}$$

$$\text{where:} \quad s = \sigma - p\mathbf{I}; \qquad \sigma = \sigma_t - \mathbf{I}p_f$$
$$\text{and:} \quad p_f = p_g \quad \text{if} \quad p_g > p_l \ ; \qquad \text{otherwise} \quad p_f = p_l \tag{A6}$$

The macrostructural bulk modulus (K_2) for changes in mean stress is evaluated with the following law:

$$K_2 = \frac{(1+e_2)}{\kappa} p \tag{A7}$$

The microstructural bulk modulus (K_1) is evaluated as follows:

$$K_1 = \frac{e^{-\alpha_m p}}{\beta_m} \tag{A8}$$

where α_m and β_m are model parameters.

The shear modulus G_t is obtained from a linear elastic model as follows:

$$G_t = \frac{3(1-2\mu)K}{2(1+\mu)} \tag{A9}$$

where μ is the Poisson's coefficient.

The macrostructural bulk modulus for changes in suction is computed considering the following law:

$$K_s = \frac{(1+e_2)(s + p_{atm})}{\kappa_s} \tag{A10}$$

where κ_s is the macrostructural elastic stiffness parameter for changes in suction.

The macrostructural bulk modulus for changes in suction is computed considering the following law:

$$K_T = \frac{1}{(\alpha_0 + \alpha_2 \Delta T)} \tag{A11}$$

where α_0 and α_2 are parameters related to the elastic thermal strain.

More details related to the model formulation and its implementation in the CODE_BRIGHT program can be found in Sánchez (2004).

A thermodynamically based model for unsaturated soils: a new framework for generalized plasticity

R. Tamagnini
University of Rome La Sapienza, Rome, Italy

M. Pastor
CEDEX, Madrid, Spain

ABSTRACT: The paper presents a generalized plasticity stress–strain relation for unsaturated soils. The constitutive equations are introduced by thermodynamics; the decomposition of the Helmholtz free energy is used to explain collapse strain recorded upon wetting. The model is formulated by a thermodynamically consistent expression of the effective stress acting on the solid skeleton. Besides, the hierarchical enhancement of the basic saturated model is obtained introducing a second plastic term. The mechanical equation is coupled with a scalar function defining the water storage mechanism which is modelled accounting for its hydraulic hysteresis. The basic saturated generalized plasticity model is recovered in the presented formulation, as special case, when the saturation degree is equal to one and the transition between saturated and unsaturated conditions takes place without discontinuity. The model can reproduce the behaviour of unsaturated soils under monotonic stress path and it is also suitable for cyclic loading. Numerical validations of the proposed equations are reported.

1 INTRODUCTION

Bishop (1959) has proposed to modify the Terzaghi's effective stress, as:

$$\sigma''_{ij} = \sigma_{ij} - u_a \delta_{ij} + \chi(u_a - u_w)\delta_{ij} \tag{1}$$

in order to model unsaturated soils behaviours. In equation 1, χ is a positive scalar function depending on S_r, σ_{ij} is the total stress, u_a and u_w are respectively the pore air pressure and the pore water pressure; in the following, suction s will be defined as the difference between these two quantities. Jennings & Bulrand (1962) have questioned that the effective stress proposed by Bishop is not able to reproduce collapse strains recorded along wetting paths. However, the authors have implicitly verified the inability of equation 1 only in the constitutive framework defined for fully saturated condition. The argumentation by Jennings and Burland has lead to the so-called *bitensorial approach*, in which constitutive equations are formulated by two independent components of the isotropic stress. Experimental validation of this approach was provided by Fredlund & Morgenstern (1977) through the results obtained in the so-called *null tests*. These tests are conducted varying mean total stress, pore air pressure and the pore water pressure maintaining suction, mean net stress (i.e. $p-u_a$) and mean effective stress (i.e. $p-u_w$) constant. As the authors have not recorded volumetric strains on samples, they have concluded that any couples of the aforementioned stress tensors are suitable for the definition of constitutive equations.

On the other hand, theoretical works based on mixture theory (Coussy 1995), (Hutter et al. 1999) averaging theory (Lewis & Schrefler 1987; Gray & Hassanizadeh 1990) or the energetic considerations by Houlsby (1997) have shown that the avereged stress acting on the solid skeleton of an unsaturated soil can be defined as:

$$\sigma_{ij}'' = \sigma_{ij} - u_a \delta_{ij} + S_r(u_a - u_w)\delta_{ij} \tag{2}$$

The stress tensor defined in equation 2 is equal to the Bishop's stress if χ is replaced by S_r.

Khogo et al. (1993), Jommi & di Prisco (1995) and Loret & Khalili (2000) have proposed constitutive models defined in terms of stress tensor equal or similar to those defined in equations 1 and 2. These formulations are able to reproduce collapse tests. These authors have used the Bishop's stress tensor in *an extended critical state framework*, where the hardening law depends even on saturation degree (or suction). Tamagnini (2004) has shown that the introduction of hydraulic hysteresis in the extended critical state framework produces a mechanical hysteresis in the hardening, that can explain the irreversible behaviour recorded during wetting-drying cycles.

These results suggest that equation 2 can be correctly adopted in the modelling of unsaturated soils. Hence, the conclusions by Jennings & Burland (1962) may be reviewed. At the same time, it should be pointed out that the experimental evidences from the null tests are consistent with the results that can be achieved by the extended critical state framework. In fact, if suction is held constant with no changes in saturation degree even the Bishop's stress tensor remains constant and the material does not harden or soften.

Gallipoli et al. (2003) and Wheeler et al. (2003) have introduced the stress tensor in equation 2 in the bitensorial approach and they argued that saturation degree must enter in the definition of the stress variables. Gens (1995) has shown that both bitensorial elasto-plastic models and constitutive equations derived by the extended critical state framework define the total strain rate as a sum of three components:

$$\dot{\varepsilon}_{ij} = \dot{\varepsilon}_{ij}^e + \dot{\varepsilon}_{ij}^p{}_\sigma + \dot{\varepsilon}_{ij}^p{}_s \tag{3}$$

where ε_{ij}^e is the elastic strain tensor, $\varepsilon_{ij}^p\sigma$ the plastic strain tensor coupled with the stress tensor containing total stress and $\varepsilon_{ij}^p s$ is the strain tensor coupled with suction (this predicts collapse). Equation 3 can be written as:

$$\dot{\varepsilon}_{ij} = \dot{\varepsilon}_{ij}^e + A_{ijkl}\dot{\sigma}_{kl}'' + B_{ij}\dot{s} \tag{4}$$

where, A_{ijkl} and B_{ij} are constitutive tensors. In literature, the following equations has been adopted for σ_{kl}'':

$$\sigma_{kl}'' = \sigma_{kl} - u_a \delta_{kl} \tag{5a}$$

$$\sigma_{kl}'' = \sigma_{kl} - u_w \delta_{kl} \tag{5b}$$

$$\sigma_{kl}'' = \sigma_{kl} - [u_a(1-S_r) + u_w]\delta_{kl} \tag{5c}$$

Stress tensor in equation 5a is used, for example, in the BBM (Basic Barcelona Model) by Alonso et al. (1990) and in its enhanced version proposed by Wheeler & Sivakumar (1995). Stress tensor in equation 5b is adopted by Geiser (1999) in the formulation of the model HiSS $\delta_{1-\text{unsat}}$. The tensor 5c is used by Jommi & di Prisco (1995) and by Tamagnini (2004) to extend the modified Cam-Clay. Wheeler et al. (2003) and Gallipoli et al. (2003) have adopted this tensor in their bitensorial models.

Even if extended critical state models and bitensorial models have the same structure (equation 3), the third terms in equation 4 differs in the two constitutive approaches. In bitensorial models the rate of suction is introduced as an external stress variable; on the other hand, in the extended critical state models this is an internal variable. This implies that bitensorial models are defined in a space of three invariants (these are generally: mean net stress, deviator stress and suction) contrary to

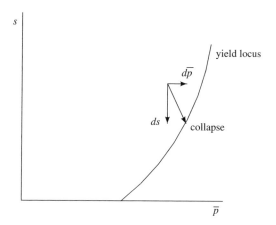

Figure 1. An ideal isotropic collapse test for constant mean Bishop stress.

the extended critical state models that are defined in the classic plane of two invariants: the mean Bishop's stress and deviator stress.

Bolzon et al. (1997) and Simoni & Schrefler (2000) have proposed a generalized plasticity model for unsaturated soils. This constitutive equation is an interesting improvement in the Bishop's monotensorial approach but it is not able to model the behaviour of soils properly along particular stress paths. The model describes soils behaviour by the following relation:

$$\dot{\varepsilon}_{ij} = \dot{\varepsilon}_{ij}^e + \frac{1}{HH(s)} \left[n_{ij} n_{gkl} \right] \dot{\sigma}_{kl}'' \qquad (6)$$

where H is the plastic modulus of the original saturated model and $H(s)$ is a scalar function depending on suction. n_{ij} and n_{gkl} are the vectors describing the loading direction and plastic flow respectively and σ_{kl}'' is the Bishop's stress. An ideal isotropic wetting test can be carried out decreasing suction and increasing the value of the mean net stress in order to maintain the mean Bishop's stress constant:

$$ds < 0 \quad \text{and} \quad d\bar{p} > 0 \quad \rightarrow \quad dp'' = 0 \qquad (7)$$

According the experimental behaviour, the ideal specimen should collapse as sketched in figure 1, on the other hand, equation 6 predicts null strains when the rate of the mean Bishop's stress is zero.

Moreover, the plasticity model defined in equation 6 does not generalise the elasto-plastic framework reported in equation 4.

Below, another hierarchical enhancement of the generalized plasticity framework is proposed. The new framework is introduced by thermodynamics. This is able to reproduce collapse strains and other typical experimental evidences recorded on unsaturated soils.

2 THERMODYNAMICS

Modern geomechanics makes use of the basic concepts of thermodynamics to evaluate constitutive assumptions. Unsaturated soils have been analysed by many authors as a multiphase mixture. These works can be divided in two groups: in the first group, the mathematical development is based on the averaging technique in which the porous media is modelled at the microscopic level and the resulting equations are then averaged on a REV (Representative Elementary Volume) to obtain macroscopic equations. The second group makes use of the mixture theory, in which the equations, such as conservation laws, are defined at the macroscale in analogy with a single phase material. Then, constitutive equations are obtained by the exploitation of the entropy inequality following the Coleman and Noll method. Another interesting approach was proposed by Gray & Hassanizadeh (1990) that is a combination of the two. Below, unsaturated soils basic equations are described following these works.

The rate of the free energy of a material during isothermal deformation can be written as:

$$d\Psi = \delta W - \delta \Phi \quad \text{with} \quad \delta \Phi \geq 0 \tag{8}$$

where, δW is the work done on the system and $\delta \Phi$ is the increment of entropy. According with the results from Coussy (1995) and Dangla et al. (1997) the work done on a three-phase mixture with two-phase flow, can be expressed in an Lagrangian form as:

$$\delta W = \sigma_{ij} \dot{\varepsilon}_{ij} + u_a \dot{v}_a + u_w \dot{v}_w \tag{9}$$

where, v_a and v_w are respectively the rate of the air volumetric content and the rate of the water volumetric content.

Introducing porosity as:

$$n = v_w + v_a \tag{10}$$

and assuming that solid grains are incompressible, volume change can be expressed as:

$$dn = -d\varepsilon_{ij} \delta_{ij} \tag{11}$$

Introducing the relations:

$$v_w = nS_r \Rightarrow dv_w = -d\varepsilon_{ij} \delta_{ij} S_r + ndS_r \tag{12}$$

$$v_a = n(1 - S_r) \Rightarrow dv_a = -d\varepsilon_{ij} \delta_{ij} + d\varepsilon_{ij} \delta_{ij} S_r - ndS_r \tag{13}$$

substituting equations 12–13 in equation 9, the incremental work results

$$\delta W = \sigma_{ij} \dot{\varepsilon}_{ij} + u_a(-\dot{\varepsilon}_{ij} \delta_{ij}) + S_r (u_a - u_w) \dot{\varepsilon}_{ij} \delta_{ij} - (u_a - u_w) n \dot{S}_r \tag{14}$$

that can be rearranged as:

$$\delta W = \left[\sigma_{ij} - u_a \delta_{ij} + S_r (u_a - u_w) \delta_{ij}\right] \dot{\varepsilon}_{ij} - (u_a - u_w) n \dot{S}_r \tag{15}$$

2.1 Thermo-elasticity

In the hypothesis of reversibility, in which dissipation doesn't occur (i.e. $\delta \Phi = 0$, equation 8 results:

$$d\Psi = \left[\sigma_{ij} - u_a \delta_{ij} + S_r (u_a - u_w) \delta_{ij}\right] d\varepsilon_{ij} - (u_a - u_w) n dS_r \tag{16}$$

that can be written as:

$$d\Psi = \sigma''_{ij} d\varepsilon_{ij} - sn dS_r \tag{17}$$

From equation 17 results:

$$\sigma''_{ij} = \frac{\partial \Psi}{\partial \varepsilon_{ij}} \tag{18a}$$

$$-ns = \frac{\partial \Psi}{\partial S_r} \qquad (18b)$$

Equation 18a shows that during a reversible elastic process the Bishop's stress tensor is the stress variable conjugated with the strains of the solid skeleton and suction scaled by porosity is the stress variable conjugated with changes in the degree of saturation. It's noteworthy that reversible changes in the degree of saturation occur only on the scanning curves joining the main drying and the main wetting curves. Equation 18b defines ns as the measure of changes in the free energy of the system produced by changes in saturation degree. Moreover, the minus of equation 18b implies that a decrease in saturation is not a spontaneous process because it requires an increase in the free energy. On the other hand, wetting can occur spontaneously.

Equation 16 can be also written as:

$$d\Psi = (\sigma_{ij} - u_a \delta_{ij})d\varepsilon_{ij} + S_r(u_a - u_w)\delta_{ij}d\varepsilon_{ij} - (u_a - u_w)ndS_r \qquad (19)$$

and for isotropic conditions:

$$d\Psi_{iso} = (p - u_a)d\varepsilon_{ii} + S_r(u_a - u_w)d\varepsilon_{ii} - (u_a - u_w)ndS_r \qquad (20)$$

from equation 20 results that elastic strain can be also conjugated with a couple of isotropic stress tensors, as adopted in the bitensorial approach. However, equation 20 suggests that suction should be multiplied by S_r. This results is obtained by Houlsby (1997).

2.2 Thermo-plasticity

Unsaturated soils behaviour is characterised by a strong hydro-mechanical coupling. This is due to the capillary forces acting within the interfacial area between fluids (i.e. meniscus water). The capillary forces are generally quantified by suction and they depend on saturation degree. Changes in suction (or in the degree of saturation) induce hydraulic and mechanical irreversible phenomena that are recorded at the macroscale. Hydraulic irreversibility is recorded during suction cycles and it causes hydraulic hysteresis in the Water Retention Curve (WRC). At the same time, a decrease in suction may induce collapse strains under certain conditions. Hassanizadeh & Gray (1993) and Dangla et al. (1997) have explained the occurrence of hydraulic hysteresis by the introduction of a second component of the free energy related to saturation degree. In this paper, the same hypothesis is adopted to explain even the collapse recorded upon wetting.

The free energy associated with the solid skeleton can be de-coupled as:

$$\delta W = d\Psi_1(\varepsilon_{ij}^e) + d\Psi_2(\varepsilon_{ij_s}^p(S_r)) + \delta \Phi \qquad (21)$$

resulting in a sum of a function of elastic strains only plus a function of only plastic strains related to the capillary forces. The de-coupling of equation 21 is a necessary and sufficient condition for the dependency of the instantaneous elastic modulus only on the elastic strains, as discussed by Lubliner (1990). Hence, equation 18a still holds even with the introduction of the hypothesis in equation 21. In particular, the free energy in equation 18a is intended as the part of the free energy related only on the elastic strains:

$$\Psi = \Psi_1(\varepsilon_{ij}^e) \quad and \quad \Psi_2(\varepsilon_{ij_s}^p(S_r)) = 0 \qquad (22)$$

The increment of plastic work is defined as:

$$\delta W^p = \sigma'_{ij}\dot{\varepsilon}_{ij}^p = d\Psi_2(\varepsilon_{ij_s}^p(S_r)) + \delta\Phi(\dot{\varepsilon}_{ij}^p) = \frac{\partial \Psi_2}{\partial \varepsilon_{ij_s}^p} \frac{\partial \varepsilon_{ij_s}^p}{\partial S_r} \frac{\partial S_r}{\partial s} \dot{s} + \delta\Phi(\dot{\varepsilon}_{ij}^p) \qquad (23)$$

or:

$$\delta\Phi(\dot{\varepsilon}_{ij}^p) = \sigma'_{ij}\dot{\varepsilon}_{ij}^p - \frac{\partial\Psi_2}{\partial\varepsilon_{ij}^p}\frac{\partial\varepsilon_{ij}^p}{\partial S_r}\frac{\partial S_r}{\partial s}\dot{s} \qquad (24)$$

where the first term on right-hand side in equation 24 represents the dissipation produced by the effective (*dissipative*) stress acting on the solid matrix and the second term is a source of the free energy depending on the capillary forces (it is assumed that the degree of saturation depends only on suction).

It is noteworthy that the second component of the free energy is a state function depending only on the current saturation degree value and it must be null on closed cycle. At the same time, it can be positive or negative provided the respect of the condition:

$$\delta\Phi(\dot{\varepsilon}_{ij}^p) = \sigma'_{ij}\dot{\varepsilon}_{ij}^p - \frac{\partial\Psi_2}{\partial\varepsilon_{ij}^p}\frac{\partial\varepsilon_{ij}^p}{\partial S_r}\frac{\partial S_r}{\partial s}\dot{s} \geq 0 \qquad (25)$$

The second term in equation 24 is a "plastic frozen work" stored during drying and loading paths in unsaturated conditions. This store of energy is due to the presence of capillary forces.

This energy can be recovered when the inter-granular bonding forces due to capillary is removed by wetting. The resulting macroscopic behaviour is called *structural collapse*. Equation 24 implies that the thermodynamic driving force of collapse is not the Bishop's stress (according figure 1) but capillary forces. Moreover, the Bishop's stress can describe completely soil's skeleton behaviour only in elasticity. Equation 24 provides a thermodynamically based interpretation of equation 4. The argumentation reported here is similar to the explanation of the kinematic hardening provided by Collins and Hilder (2002). A detailed discussion about this topic will appear on (Tamagnini, in prep.).

3 A GENERALIZED PLASTICITY MODEL FOR UNSATURATED SOILS

3.1 *Saturated soils*

Constitutive equations for saturated soils can be introduced without defining any yield and potential surface as proposed by Pastor et al. (1990). Constitutive equations for saturated soils are written as:

$$\dot{\varepsilon}_{ij} = \dot{\varepsilon}_{ij}^e + \frac{1}{H_{L/U}}[n_{ij}n_{gkl}]\dot{\sigma}'_{kl} \qquad (26)$$

where σ'_{kl} is the Terzaghi or Biot's effective stress tensor, $H_{L/U}$ is the plastic modulus defined for loading conditions as:

$$H_L = H_0 p' H_f \{H_v + H_s\} H_{dm} \qquad (27)$$

and during unloading:

$$H_U = \infty \text{ (for clays)} \qquad (28)$$

$$H_U = H_{U0}\left(\frac{M_g}{\eta_u}\right)^{\gamma_u} \quad \text{for } \left|\frac{M_g}{\eta_u}\right| > 1$$

$$H_U = H_{U0} \quad \text{for } \left|\frac{M_g}{\eta_u}\right| \leq 1 \qquad \text{(for sands)} \qquad (29)$$

where H_0 defines plastic modulus during isotropic compression, p' is the isotropic effective stress, H_f accounts for the loading direction, H_s accounts for deviator strain hardening and H_{dm} for the memory of the past stress history. H_U is the plastic modulus defining plastic strains during unloading. Explicit expressions for clays and sands are given by Pastor et al. (1990).

3.2 Unsaturated soils

The hierarchical enhancement of the equation 26, as proposed by Tamagnini (2003), is:
Mechanical equation:

$$\dot{\varepsilon}_{ij} = C^e_{ijkl}\dot{\sigma}''_{kl} + \frac{1}{H_0\sigma''_{rr}\{H_v + H_s\}H_{dm}}[n_{gij}n_{kl}]\dot{\sigma}''_{kl} + \frac{1}{H_b}n_{gij}\dot{s} \qquad (30)$$

Hydraulic equations:

$$S_{rD/W} = WRC(ns) \qquad (31)$$

$$S_r = k(ns) \qquad (32)$$

Equations 31–32 define the hydraulic constitutive relationship describing the water-storage mechanism. Equation 31 can be specified to define the Main drying and Main wetting curves; equation 33 defines the scanning curves. Equations 31 and 32 can reproduce hydraulic hysteresis. Tamagnini (2004) has shown that the introduction of hysteresis in the hydraulic part of elasto-plastic models allows for reproducing inelastic behaviour recorded during cyclic wetting-drying tests. In equation 30 σ''_{kl} is the stress tensor defined in equation 18a. The terms in the hardening modulus and the vectors n_{kl} and n_{gij} are redefined accounting for the new stress ratio:

$$\eta = \frac{q}{p''} \qquad (33)$$

where p'' is the isotropic Bishop's stress and q is the deviator stress. The second component of plastic strain in equation 30 is introduced according with the general framework described by equation 4 and it can be interpreted by the thermodynamics of equation 24. In equation 30, the original H_{dm} is modified to account for the double mechanism of hardening induced by both suction and plastic strain:

$$H_{dm} = \left(\frac{\zeta_{MAX} J(s)}{\zeta}\right)^\gamma \qquad (34)$$

ζ is the mobilized stress function defined as:

$$\zeta = p''\left\{1 - \left(\frac{1}{1+\alpha}\right)\frac{\eta}{M}\right\}^{-\frac{1}{\alpha}} \qquad (35)$$

equation 35 is the original expression for the saturated material modified expressed in terms of the Bishop's stress. ζ_{MAX} is the maximum value previously reached by function ζ. The equation J provides the additional form of hardening due to partial saturation and it is stated as:

$$J = \exp[c(1 - S_r)] \qquad (36)$$

where, c is a constitutive parameter. In the presented model ζ_{MAX} is updated only when ζ satisfies the condition $\zeta \geq \zeta_{MAX} J$.

The plastic modulus in the third term of equation 30 defines plastic strain produced by changes in suction during wetting and it is stated as:

$$H_b = wH_0 p'' H_f H_{dm} \qquad (37)$$

The modulus of equation 37 can be determined starting from a wetting path in which the material undergoes collapse. Wetting path is an unloading stress path, according with the definition of the adopted Bishop's stress. Hence, during a wetting path, the second dissipative term of equation 30 is imposed as zero. This condition agrees with the formulation of the original saturated model (see equation 28). This feature is another remarkable difference with the use of the generalized plasticity proposed Bolzon et al. (1996). During unloading, equation 30 results:

$$\dot{\varepsilon}_{ij} = C^e_{ijkl}\dot{\sigma}''_{kl} + \frac{1}{H_b}n_{g\,ij}\dot{s} \qquad (38)$$

The modulus H_b is obtained hierarchically starting from isotropic condition. Isotropic collapse is stated as proportional (through the negative constitutive parameter w) to the plastic strain occurring in isotropic strain hardening:

$$\dot{\varepsilon}^p_v = \frac{1}{wH_o p''H_{dm}}\dot{s} \qquad (39)$$

w can be constant or it can be assumed as to be function of s.

Equation 38 predicts *collapse* if the second term prevails on the first. This depends on the modulus in equation 37. Particularly, if the "over-consolidation" produced by both strain hardening and "bonding" exerted by suction is large the resulting plastic component of equation 38 is negligible with respect to the elastic expansion predicted by the first term. On the contrary, if H_{dm} is almost one plastic compression occurs. For isotropic stress condition equation 35 is:

$$\zeta = p'' \qquad (40)$$

Plastic compression is small if:

$$\zeta \ll \zeta_{MAX} J(s) \qquad (41)$$

The evolution of the elastic and plastic isotropic strains are depicted in figure 2. This figure shows that collapse strain grows when the applied Bishop's stress increases and, for a fixed suction value, when total stress increases.

The influence of the loading direction on collapse strain is not well experimentally investigated. In the presented formulation H_f is defined as proposed by Pastor et al. (1990):

$$H_f = \left(1 - \frac{\eta}{M_g}\right)^\mu \frac{(1+d_0^2)}{(1+d^2)}\left|1 - \frac{\eta}{M_g}\right| \qquad (42)$$

where $d = (1+\alpha)(M-\eta)$ and α is a constitutive parameters. The introduction of equation 42 in plastic modulus H_b implies a maximum collapse in isotropic condition and null collapse strain at the critical state. M_g is the slope of the critical state line.

The direction of the plastic flow produced during collapse coincides with the plastic flow $\mathbf{n}_g$ produced by loading. This assumption agrees, with the experimental results from Rifa'I et al. (2002).

The modulus H_v is defined as:

$$H_v = \left(1 - \frac{\eta}{M_g}\right) \qquad (43)$$

The modulus H_s is:

$$H_s = \beta_0\beta_1 \exp(-\beta_0\xi) \qquad (44)$$

β_1 and β_0 are constituve parameters and ξ is the amount of accumulated deviator plastic strain.

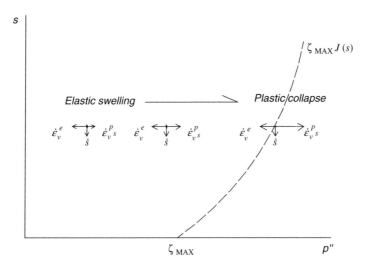

Figure 2. Evolution of the two strain components of equation 38 in isotropic wetting path for different value of the H_{dm}.

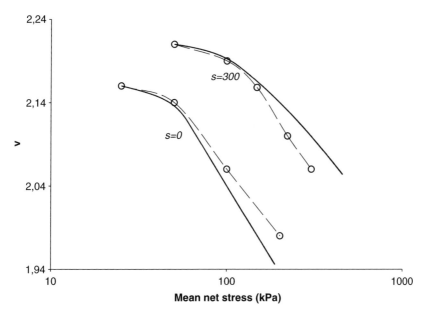

Figure 3. Comparison of predicted and experimental data (after Wheeler & Sivakumar (1995)) during isotropic compression at constant suction.

4 VALIDATION

4.1 *Constant suction*

Figure 3 reports the simulation (solid lines) of the laboratory data (dots) from Wheleer & Sivakumar (1995) obtained during isotropic compressions at constant suction of 0 and 300 kPa. Parameters of equation 30 are: $\lambda = 0.15, \kappa = 0.02, \gamma = 5.0, v_0 = 2.16$ and $p'_0 = 25$ (corresponding to a value of

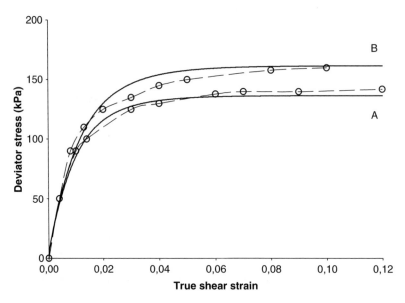

Figure 4. Comparison of predicted and experimental data (after Wheeler & Sivakumar 1995) for constant volume and constant suction shear tests.

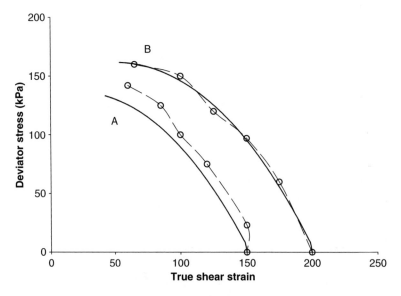

Figure 5. Comparison of predicted and experimental data (after Wheeler & Sivakumar 1995) for constant volume and constant suction shear tests.

$H_{dm} = 1.8$) for the compression at $s = 0$ and $\lambda = 0.15, \kappa = 0.02$ kPa, $\gamma = 5.0, v_0 = 2.21, p'_0 = 126$ (corresponding to a value of $H_{dm} = 1.8$) for the compression at $s = 300$.

A comparison between the experimental data (dots) from (Wheleer & Sivakumar 1995) and the model prediction (solid lines) for a constant volume and constant suction test is reported in figure 4 and figure 5. A comparison between the experimental data (dots) from (Wheleer & Sivakumar 1995) and the model prediction (solid lines) for a constant mean net stress and constant suction test is reported in figure 6 and figure 7.

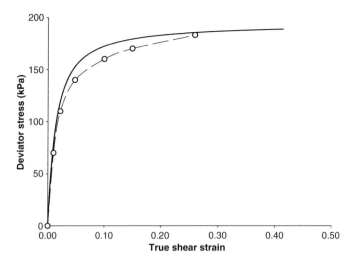

Figure 6. Comparison of predicted and experimental data (after Wheeler & Sivakumar 1995) for constant mean net stress and constant suction shear tests.

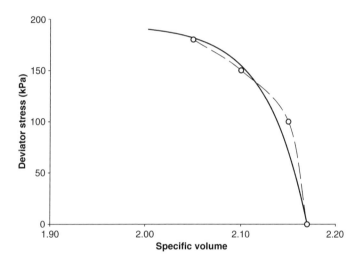

Figure 7. Comparison of predicted and experimental data (after Wheeler & Sivakumar 1995) for constant mean net stress and constant suction shear tests.

4.2 Collapse tests

Figure 8 reports the comparison of the predicted behaviour and experimental data of two collapse/swelling tets from (Escario & Saez 1973). Samples are wetted starting from the value of suction of 3500 kPa and 1500 kPa. The model is characterized by the parameters: $\lambda = 0.105, \kappa = 0.04, \gamma = 5.0, v_0 = 2.3, H_{dm} = 1.0, c = 2.5$.

5 A NEW FRAMEWORK FOR GENERALIZED PLASTICIY

Different soils behaviours can be modelled by the same constitutive assumptions. Thermal effects, collapse in unsaturated soils and rockfill or the effects of chemical degradation (Nova et al. 2003) can be modelled by a Cam-clay family model in which the hardening rule is extended and the

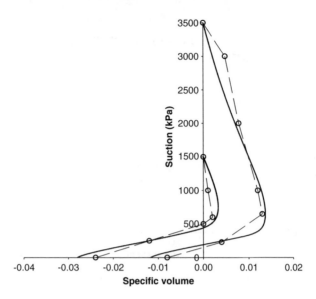

Figure 8. Comparison of predicted and experimental data (Escario and Saez 1973) for swelling collapse tests.

effective stress is redefined. These mathematical modifications do not depend on the particular constitutive laws for the unbonded, isothermal saturated material. Hence, these can be hierarchically added to each classic elastic-plastic models and these can be generalized as follow:

Yield locus:

$$f(\sigma_{ij}; h(\varepsilon_{ij}^p; X)) = 0 \tag{45}$$

Plastic potential:

$$g(\sigma_{ij}; h(\varepsilon_{ij}^p; X)) = 0 \tag{46}$$

Hardening rule:

$$h = h(\varepsilon_{ij}^p; X) \tag{47}$$

where X is a second independent internal variable affecting the hardening. It can be substituted by temparature, saturation degree, suction or solvent concentration in order to model collapse.

Plastic flow:

$$\dot{\varepsilon}_{ij}^p = \dot{\lambda} \frac{\partial g}{\partial \sigma_{ij}} \tag{48}$$

Consistency:

$$\dot{f} = \frac{\partial f}{\partial \sigma_{ij}} \dot{\sigma}_{ij} + \frac{\partial f}{\partial h} \frac{\partial h}{\partial \varepsilon_{ml}^p} \dot{\varepsilon}_{ml}^p + \frac{\partial f}{\partial h} \frac{\partial h}{\partial X} \dot{X} = 0 \tag{49}$$

Plastic multiplier:

$$\dot{\lambda} = \frac{1}{H} \frac{\partial f}{\partial \sigma_{ij}} \dot{\sigma}_{ij} + \frac{1}{H_X} \dot{X} \tag{50}$$

in which:

$$H_X = H\left(\frac{\partial f}{\partial h}\frac{\partial h}{\partial X}\right)^{-1} \quad (51)$$

$$H = -\frac{\partial g}{\partial \sigma_{lm}}\frac{\partial f}{\partial h}\frac{\partial h}{\partial \varepsilon^p_{lm}} \quad (52)$$

Constitutive equation:

$$\dot{\varepsilon}_{ij} = C^e_{ijkl}\dot{\sigma}_{kl} + \frac{1}{H}\frac{\partial f}{\partial \sigma_{ij}}\frac{\partial g}{\partial \sigma_{lm}}\dot{\sigma}_{lm} + \frac{1}{H_X}\frac{\partial g}{\partial \sigma_{lm}}\dot{X} \quad (53)$$

equation 53 is generally coupled with a scalar constitutive equation defining the evolution of X. For example, in unsaturated soils this scalar equation is the WRC.

The second term of equation 53 represents the plastic strain that is produced to respect the persistency condition when the rate of X drives the current stress outside of the elastic domain and it computes the degradation, collapse upon wetting or thermal softening.

Even if the rate of X is contained in the rate of the stress σ_{lm} the two plastic terms can not be added, because plastic strain (collapse, chemical degradation, etc.) can occur even during an unloading (or neutral load). Mathematically:

$$\frac{\partial f}{\partial \sigma_{ij}}\dot{\sigma}_{ij} \leq 0$$

$$\frac{1}{H}\frac{\partial g}{\partial \sigma_{ij}}\frac{\partial f}{\partial \sigma_{lm}}\dot{\sigma}_{lm} = 0 \quad (54)$$

$$\frac{1}{H_X}\frac{\partial g}{\partial \sigma_{ij}}\dot{X} = \varepsilon^p_{ij} \quad \text{(collapse strain)}$$

Condition 54 explains because the so-called *Bishop's approach* (and not the Bishop's stress) is inconsistent with the experimental behaviour of collapsible soils. The Bishop's approach disregards the influence of suction on the hardening and then it does not define the third term of equation 53.

6 CONCLUSION

A generalized plasticity framework for unsaturated soils has been presented. The de-coupling in the free energy has been introduced to explain the structure of the constitutive equations and the physics of collapse strains. It has been shown that even if the Bishop's stress is the thermodynamically consistent expression of the stress tensor, this is not the thermodynamic driving force of collapse. The proposed model has been applied in the prediction of soils behaviours under monotic stress path but it is also a suitable formulation to predict cyclic response, such occurs in earthquakes or cyclic changes in water content. Moreover, It has been shown as the new generalised framework may be also adopted to model other degradation phenomena.

ACKNOWLEDGMENTS

Prof. C. Jommi is acknowledged for the help and for the useful discussions.

REFERENCES

Alonso, E.E, Gens, A. and Josa, A. 1990. A constitutive model for partially saturated soils. *Geotechnique* 40: 405–430

Bishop, A.W. 1959. The principles of effective stress, *Tecknisk Ukeblad*, 106(39): 859–863

Bolzon, G., Schrefler, B.A. & Zienkiewicz, O.C., 1996. Elastoplastic soil constitutive laws generalized to partially saturated state. *Geotechnique* 46(2): 279–289

Collins, I.F. and Hilder, T. (2002) A theoretical framework for constructing elastic/plastic constitutive models of triaxial tests, Int. J. for Num. and Anal. Meth. in Geomech. V. 26(13):1313–1347

Coussy, O. 1995 *Mechanics of porous continua*, New York: J. Weley and Sons

Dangla P., Malinski, L. & Coussy, O. 1997. Plasticity and imbibition –drainage curves for unsaturated soils: A unified approach. In Petruzckzacs & Pande (eds) *Proc. VII Int. Symposium on Numerical Models in Geomechanics*, Montreal, 2–4 July: 141–146

Escario, V. and Saez, J., 1973. Measurement of properties of swelling and collapsing soils under controlled suction. *In Proc. 3rd Int. Conf. Expansive Soils*, Haifa: 196–200

Fredlund, D.G. & Morgenstern, N.R. 1977. Stress state variables for unsaturated soils. *J. Geot. Eng. Div. ASCE* 103(GT5): 447–466

Gallipoli, D. Gens, A. Sharma, R. & Vaunat, J. 2003. An elasto-plastic model for unsaturated soil incorporating the effects of suction and degree of saturation on mechanical behaviour. *Geotechnique*. 53(1): 123–135.

Geiser, F. 1999 Comportement mécanique d'un limon non saturé. PhD Thesis. Losanne

Gens, A. 1995 Costitutive modelling: Application to compacted soils In Alonso & Delage (eds) *Proc. of the first Int. Conf. on Unsaturated Soils*/UNSAT 95/*Paris*: 1179–1200

Gray, G.G. & Hassanizadeh, S.M. 1990. Mechanics and thermodynamics of multiphase flow in in porous media including interphase boundaries. *Adv. Water Resources* 13(4): 169–186

Kohgo Y., Nakano M. & Miyazaki, T. (1993) Theoretical Aspects of Constitutive Modelling for Unsaturated Soils, *Soils and Foundations*, Vol. 33, No. 4: 49–63

Lubliner J. 1990 *Plasticity theory* New York: McMillan

Loret, B. & Khalili, N. 2000 A three-phase model for unsaturated soils. *International Journal for Numerical and Analytical Methods in Geomechanics* 24(11): 893–927

Lewis, R.W. & Schrefler, B.A. 1987 *The Finite Element Method in the Deformation and Consolidation of Porous Media*: Chichester J. Wiley and Sons

Hassanizadeh, S.M. & Gray, G. G. 1993. Thermodynamic basis of capillary in porous media. *Water Resources Research*. 29(10): 3389–3405

Houlsby, G.T. 1997. The work input to an unsaturated granular material. *Geotechnique* 47(1): 193–196

Hutter, K., Laloui, L. & Vulliet, L. 1999. Thermodynamically based mixtures models of saturated and unsaturated soils. *Mech. Cohes.-Frict. Mat.*4: 295–338

Jennings, J.E.B. & Burland, J.B. 1962. Limitation to the use of effective stress in partly saturated soils., *Geotechniqu*e 12(2): 125–144

Jommi C. & di Prisco C. 1994. A simple theoretical approach for modelling the mechanical behaviour of unsaturated soils (in Italian) *In Conf. Il ruolo dei fluidi nei problemi di ingegneria geotecnica*, Mondovi: 167–188

Nova R., Castellanza R. & Tamagnini C. (2003) A constitutive model for geomaterials subject to mechanical and/or chemical degradation, *Int. J. Anal. Num. Methods in Geomech.*, Vol. 27(9): 705–732

Pastor, M., Zienkiewicz, O.C. & Chan, A.H.C., 1990. Generalzed plasticity and the modelling of soils behaviour. *Int. J. Num. Anal. Meth. Geomech,* 14: 151–190

Rifa'i, A., Laloui, L., Deschamps, D. & Vulliet, L. (2002) Effect of wetting process on the yield limit of a remoulded silt, Unsaturated Soils. In Jucá., de Campos, and Marinho (eds), *Proc. 3rd Int. Conf. on Unsaturated Soils (UNSAT 2002)*, Recife, Brazil, Vol. 1: 159–166

Simoni, L. & Schrefler, B. A., 2001. Parameter identification for a suction-dent plasticity model. *Int. J. Num. Anal. Meth. Geomech.* 25: 273–288

Tamagnini, R. 2003. *Analytical and Numerical models for unsaturated soils*. Rome: PhD Thesis

Tamagnini, R. 2004. An extended Cam-clay model for unsaturated soils with hydraulic hysteresis. *Geotechnique* 54(3): 223–228

Wheeler, S.J. & Sivakumar, V., 1995. An Elasto-Plastic Critical State Framework for Unsaturated Soil, *Geotechnique* 45(1): 35–53

Wheeler, S.J., Sharma, R.S. & Buisson, M. S. R. 2003. Coupling of hysteresis and stress-strain behaviour in unsaturated soil, *Géotechnique* 53(1): 41–54

Miscellaneous

Opening lecture

Prof. Eduardo Alonso – Universitat Politècnica de Catalunya, Barcelona, Spain

Discussion leaders

Stefano Aversa – Università di Napoli "Parthenope", Italy
Pierre Delage – Ecole Nationale des Ponts et Chaussées, Paris, France
Antonio Gens – Universitat Politècnica de Catalunya, Barcelona, Spain
Luigi Mongiovì – Università di Trento, Italy
Simon Wheeler – University of Glasgow, United Kingdom

Author addresses

Marta Boso
bosom@ing.unitn.it
Dipartimento di Ingegneria Meccanica e Strutturale
Università degli Studi di Trento
Via Mesiano 77
38050 Trento, Italy

Carlos Buenfil
carlos.buenfil@upc.es
Departament d'Enginyeria del Terreny, Cartogràfica i Geofísica
Universitat Politècnica de Catalunya
C. Jordi Girona 1-3, Edificio D2
08034 Barcelona, Spain

Elisabetta Cattoni
cattonie@unipg.it
Dipartimento di Ingegneria Civile e Ambientale
Università degli Studi di Perugia
Via Duranti 93
06125 Perugia, Italy

Azad Koliji
azad.koliji@epfl.ch
Soil Mechanics Laboratory (LMS)
Swiss Federal Institute of Technology Lausanne
EPFL – Ecublens
1015 Lausanne, Switzerland

Yasar Kamil Kazimoglu
y.kazimoglu@napier.ac.uk
School of Built Environment
Napier University
10 Colinton Road
Edinburgh EH105DT, United Kingdom

Grégoire Priol
priol@cermes.enpc.fr
Centre d'Enseignement et de Recherche en Mécanique des Sols (CERMES)
Ecole Nationale des Ponts et Chaussées
6–8 Av. Blaise Pascal
Cité Descartes Champs-sur-Marne
77455 Marne-La-Vallée Cedex 2, France

Adrian Russel
A.Russell@bristol.ac.uk
Department of Civil Engineering
University of Bristol
Room 2.37, Queen's Building
Bristol BS81TR, United Kingdom

Marcelo Sánchez
marcelo.sanchez@upc.es
Departament d'Enginyeria del Terreny, Cartogràfica i Geofísica
Universitat Politècnica de Catalunya
C. Jordi Girona 1-3, Edificio D2
08034 Barcelona, Spain

Roberto Tamagnini
sgi_tamagnini@studio-geotecnico.it
Studio Geotecnico Italiano
Via Ripamonti, 89
20141 Milano, Italy

Lúcio Flávio de Souza Villar
lvillar@etg.ufmg.br
Departamento de Engenharia de Transportes e Geotecnia
Universidade Federal de Minas Gerais
Avenida do Contorno, 842/608 – Centro
30110-060 – Belo Horizonte – MG, Brazil

List of participants

Andrea Zenga	Università di Roma, La Sapienza, Italy
Camillo Airò Farulla	Università di Palermo, Italy
Eduardo Alonso	Universitat Politècnica de Catalunya, Spain
Stefano Aversa	Università di Napoli Partenope, Italy
Renato Belviso	Politecnico di Bari, Italy
Marta Boso	Università di Trento, Italy
Alessandro Buscemi	Università di Trento, Italy
Francesco Cafaro	Politecnico di Bari, Italy
Paola Caporaletti	Università di Roma La Sapienza, Italy
Marco Caruso	Politecnico di Milano, Italy
Francesca Casini	Università di Roma La Sapienza, Italy
Elisabetta Cattoni	Università di Perugia, Italy
Manuela Cecconi	Università di Perugia, Italy
Renato Maria Casentini	Università della Calabria, Italy
Tacio De Campos	Pontifìcia Universidade Catòlica do Rio de Janeiro, Brazil
Pierre Delage	Ecole Nationale des Ponts Chaussées, France
Anna D'Onofrio	Università di Napoli Federico II, Italy
Francesca D'Onza	Università di Napoli Federico II, Italy
Aldo Evangelista	Università di Napoli Federico II, Italy
Alessio Ferrari	Università di Palermo, Italy
Enzo Fontanella	Università di Roma La Sapienza, Italy
Antonio Gens	Universitat Politècnica de Catalunya, Spain
Vladislava Herbstova	Charles University, Prague, Czech Republic
Cristina Jommi	Politecnico di Milano, Italy
Yasar Kamil Kazimoglu	Napier University, UK
Azad Koliji	Swiss Federal Institute of Technology at Lausanne, Switzerland
Claudio Mancuso	Università di Napoli Federico II, Italy
Miguel Angel Martin	Universitad Politécnica de Madrid, Spain
John McDougall	Napier University, UK
Luigi Mongiovì	Università di Trento, Italy
Marco Valerio Nicotera	Università di Napoli Federico II, Italy

Luca Pagano	Università di Napoli Federico II, Italy
Vincenzo Pane	Università di Perugia, Italy
Angelina Parlato	Università di Napoli Federico II, Italy
Grégoire Priol	Ecole Nationale des Ponts Chaussées, France
Cristina Rabozzi	Politecnico di Torino, Italy
Enrique Romero	Universitat Politècnica de Catalunya, Spain
Adrian Russel	University of Bristol, UK
Giacomo Russo	Università di Cassino, Italy
Marcelo Sánchez	Universitat Politècnica de Catalunya, Spain
Filippo Santucci De Magistris	Università di Napoli Federico II, Italy
Anna Scotto di Santolo	Università di Napoli Federico II, Italy
Stefania Sica	Università di Napoli Federico II, Italy
Giuseppe Sorbino	Università di Salerno, Italy
Roberto Tamagnini	Università di Roma La Sapienza, Italy
Alessandro Tarantino	Università di Trento, Italy
Valerio Tedesco	Università di Cassino, Italy
Sara Tombolato	Università di Trento, Italy
Roberto Vassallo	Università della Basilicata, Italy
Sara Vecchietti	Università di Perugia, Italy
Lúcio Flávio de Souza Villar	Universidade Federal de Minas Gerais, Belo Horizonte, Brazil
Simon Wheeler	University of Glasgow, UK
Maria Claudia Zingariello	Università di Napoli Federico II, Italy

Author index

Boso, M. 1
Buenfil, C. 15

Campos de, T.M.P. 55
Cattoni, E. 29
Cecconi, M. 29
Cui, Y.-J. 43
Cuisinier, O. 83

Delage, P. 43

Gennaro De, V. 43
Gens, A. 15, 107

Kazimoglu, Y.K. 71
Khalili, N. 95
Koliji, A. 83

Laloui, L. 83
Lloret, A. 15

McDougall, J.R. 71
Mongiovì, L. 1

Olivella, S. 107

Pane, V. 29
Pastor, M. 121

Priol, G. 43
Pyrah, I.C. 71

Romero, E. 15
Russell, A.R. 95

Sánchez, M. 107
Souza Villar de, L.F. 55

Tamagnini, R. 121
Tarantino, A. 1

Vulliet, L. 83